OUT OF THE EARTH

OUT OF THE EARTH

CIVILIZATION AND THE LIFE OF THE SOIL

DANIEL J. HILLEL

THE FREE PRESS · *A Division of Macmillan, Inc.* · *New York*
COLLIER MACMILLAN CANADA · *Toronto*
MAXWELL MACMILLAN INTERNATIONAL · *New York* · *Oxford* · *Singapore* · *Sydney*

The Free Press
A Division of Macmillan, Inc.
866 Third Avenue, New York, N.Y. 10022

Collier Macmillan Canada, Inc.
1200 Eglinton Avenue East
Suite 200
Don Mills, Ontario M3C 3N1

Printed in the United States of America

printing number
1 2 3 4 5 6 7 8 9 10

Library of Congress Cataloging-in-Publication Data

Hillel, Daniel.
    Out of the earth : civilization and the life of the soil / Daniel
Hillel.
        p.   cm.
    Includes bibliographical references and index.
    ISBN 0-02-915060-4
    1. Soils.   2. Soil and civilization.   3. Water and civilization.   4. Water—
supply.   5. Agriculture.   I. Title.
S591.H62   1990
631.4—dc20
                                                                                    90-38119
                                                                                    CIP

For my daughter Shira, who was with me throughout
the writing of this book; and
for my sisters Helen, Evelyn, Bernice, and Lolly
who have always been with me

CONTENTS

ACKNOWLEDGMENTS

A FEW YEARS AGO I was invited to deliver a public lecture to the faculty and students of the University of Massachusetts on the basic principles and current issues of my profession of soil and water science, as a vital aspect of environmental science. The honor entailed the challenge of presenting the essentials of that profession to the educated lay public in a way that would be succinct and interesting, yet not superficial. The challenge stayed with me long after that lecture was delivered, and it impelled me to undertake the larger effort that has culminated at last in this manuscript. I wish therefore to express my gratitude first to the University of Massachusetts, my academic home for the last thirteen years, for granting me the opportunity to express what had been brewing in my mind for a very long time and to pursue my work—on this and other research topics—in an atmosphere of true academic freedom.

On a more personal level, I am grateful to Michal Artzy for helping me to find and interpret some of the historical and archaeological evidence pertaining to the uses and abuses of soil and water over the course of civilization; to David Hopper, erstwhile senior vice-president of the World Bank, for taking x an early interest in my manuscript and for sharing his global perspective with me. I also wish to thank Ed Rothstein, former acquiring editor at The Free Press; my present editor, Adam Bellow; Victor Rangel-Ribeiro, copy editor; Edith Lewis, editing supervisor; and Erwin Glikes, President and Publisher of The Free Press, for their special interest and encouragement.

Finally, I owe the greatest debt of gratitude to my teachers for awakening my interest in the science of the environment and the life of the soil, and for imbuing me with their love of the land. Among my early mentors were Professors J. S. Joffe, F. E. Bear, and S. A. Waksman of Rutgers University, outstanding scientists who were also inspired and inspiring philosophers; Professors S. Ravikovitch and A. Reifenberg of the Hebrew University, who combined science with a profound perception of history and archaeology; and in particular Dr. W. C. Lowdermilk, a founder of the U.S. Soil Conservation Service and early advisor to the Israel Ministry of Agriculture, a true pioneer of land and water conservation. I have subsequently been influenced by many colleagues and students from all over the world. I hope I have not failed them in my attempt to convey the special fascination and the universal importance of the quest to which we have devoted our careers.

PART I

FOR

SOIL

THOU

ART

*We know more about the movement
of celestial bodies than about the
soil underfoot.*

LEONARDO DA VINCI

1

PROLOGUE

ALL TERRESTRIAL LIFE ultimately depends on soil and water.
So commonplace and seemingly abundant are these elements that we tend to treat them contemptuously. The
very manner in which we use such terms as "dirty," "soiled,"
"muddled," and "watered down" betrays our disdain. But, in
denigrating and degrading these precious resources, we do ourselves and our descendants great—and perhaps irreparable—
harm, as shown by the disastrous failures of past civilizations.

Before I began my research, I had held the rather prevalent
idea that human abuse of the environment is a new phenomenon,
mostly a consequence of the recent population explosion and of
our expansive modern technological and materialistic economy.
Ancient societies, I presumed, were more prudent than ours in
the way they treated their resources. For the most part, that
has turned out to be a romantic fiction. My research has led
me to the conclusion that manipulation and modification of
the environment was a characteristic of many societies from their
very inception. Long before the advent of earth-moving machines
and toxic chemicals, even before the advent of agriculture, humans

3

began to affect their environment in far-reaching ways that destabilized natural ecosystems.

In many of the older countries, where human exploitation of the land began early in history, we find shocking examples of once-thriving regions reduced to desolation by man-induced soil degradation. Some of these civilizations succeeded all too well at first, only to set the stage for their own eventual demise. Consider, for example, the southern part of Mesopotamia ("the land between the rivers") which, as every schoolchild knows, was a great "cradle of civilization." We need only fly over this ancient country, now part of Iraq, to observe wide stretches of barren, salt-encrusted terrain, crisscrossed with remnants of ancient irrigation canals. Long ago, these were fruitful fields and orchards, tended by enterprising irrigators whose very success inadvertently doomed their own land.

The poor condition of the "Fertile Crescent" today is due not simply to changing climate or to the devastations caused by repeated wars, though both of these may well have had important effects. It is due in large part to the prolonged exploitation of this fragile environment by generations of forest cutters and burners, grazers, cultivators, and irrigators, all diligent and well intentioned but destructive nonetheless. The once-prosperous cities of Mesopotamia are now *tells,* mute time capsules in which the material remnants of a civilization that lived and died there are entombed. Similarly ill-fated was the ancient civilization of the Indus Valley in present-day Pakistan.

A haunting example of soil abuse on a large scale can be seen in the Mediterranean region, which has borne the brunt of human activity more intensively and for a longer period than any other region on earth. Visit the hills of Israel, Lebanon, Greece, Cyprus, Crete, Italy, Sicily, Tunisia, and eastern Spain. There, rainfed farming and grazing were practiced for many centuries on sloping terrain, without effective soil conservation. The land had been denuded of its natural vegetative cover, and the original mantle of fertile soil, perhaps one meter deep, was raked off by the rains and swept down the valleys toward the sea. That may have been the reason why the Phoenicians, Greeks,

Carthagenians, and Romans, each in turn, were compelled to venture away from their own country and to establish far-flung colonies in pursuit of new productive land. The end came for each of these empires when it had become so dependent on faraway and unstable sources of supply that it could no longer maintain central control.

The inability to ensure a dependable supply of water has also been a frequent cause of failure. A poignant example is the sad fate of Fatehpur Sikri, the magnificent capital built in northern India in the late sixteenth century by the Moghul emperor, Akbar the Great. Less than two decades after its completion, notwithstanding the splendor of its architecture, Fatehpur Sikri was abandoned entirely, for no other reason than the simple lack of water. Still more significant were the chain-well systems developed in ancient Persia. Some of these have remained in operation for several millennia, while abandoned remnants of others stand as mute testimony to the dangers of groundwater mismanagement.

There were, on the other hand, a few societies that did better than others. Some ingenious and diligent societies developed technologies that enabled them to thrive in difficult circumstances for many centuries. Judicious management of soil and water is exemplified in some of the arid regions of the Near East and the American Southwest. Equally impressive is the evidence regarding the long-lasting wetlands-based societies of Meso-America and South America. Remarkably productive wetland management systems have survived intact in China and other parts of Southeast Asia. In contrast with the historic failures of Mesopotamia and the Indus Valley, the irrigation-based civilization of Egypt sustained itself for more than five millennia— though it is now beset with problems of unprecedented severity.

Every one of the insidious man-induced scourges that played so crucial a role in the decline of past civilizations has its mirror image in our contemporary world. But it seems that the mirror is warped, and the problems it reflects are magnified and made monstrously grotesque. Human treatment of the environment has grown worse, and in our generation it has brought us to a point of crisis. Salinization, erosion, denudation of watersheds, silting of valleys and estuaries, degradation of arid lands, depletion and pollution of water resources, abuse of wetlands, and excessive population pressure—all are now occurring more intensively and on an ever-larger scale. Added to the old problems are entirely new ones, including pesticide and fertilizer residues, domestic and industrial wastes, the poisoning of groundwater, air pollution and acid rain, the mass extinction of species and, finally, the threat of global climate change.

Among the most egregious examples of latter-day abuse is the drying of the Aral Sea in the USSR, once the world's fourth largest fresh-water lake, now made briny and charged with poisonous chemical residues. An even greater disaster is the progressive decimation of the tropical rain forests and the resulting wholesale eradication of entire ecosystems. Intensified runoff, accelerated erosion, and flooding of lowlands are now widespread, and in places—for example, in Bangladesh—the results are disastrous. The degradation of vegetation and land in arid regions, a process called desertification, is occurring on a continental scale in Africa and elsewhere. Irrigated lands in such disparate countries as Australia, Pakistan, India, USSR, and the United States are losing their initially bountiful fertility and in district after district are being withdrawn from production.

Yet there are hopeful developments, too. We know much more about the natural and man-induced processes at work; we understand and can anticipate some of their consequences. Degradation and pollution are not inevitable. They can be controlled. We can avoid the major abuses and devise better modes of environmental management. Land and water husbandry can be improved and sustained.

As the reader may have already noticed, this book is not a strictly dispassionate exposition. Its topic is of intense personal interest to me. That interest has been lifelong. Born in a man-made oasis in the semi-desert of southern California, I was taken at an early age to Palestine, then in the first stages of reclamation from centuries of desolation. I spent part of my childhood in a pioneering settlement in the Jezreel Valley where, in Biblical times, Gideon drove off the hordes of desert nomads who periodically descended like locusts upon the laboriously cultivated fields of settled farmers.

Here, the ancient struggle between the wilderness and the sown, between wandering herders and sedentary cultivators, between the descendants of Abel and Cain, has been waged since civilization began. And it was here at the thin edge of life that I was first captivated by the land and its contrasts, indeed by the whole environmental symphony with its counterpoints of sky and earth, soil and water, plants and animals, wilderness and agriculture. I remember myself as a child of nine, standing barefoot in an irrigated furrow, mired in the squishy ooze, gazing at the gurgling waters slaking the harsh dry clods, marveling at the exuberant growth of tender saplings in the tiny watered grove that rose up so defiantly against the vast expanse of the surrounding arid plain.

That early fascination with soil and water has grown over the decades to become both a vocation and an avocation, a professional career and a labor of love. After earning academic degrees in agronomy and the earth sciences in America, and working for the U.S. Department of Agriculture, I returned to Israel shortly after its birth as a state. There I took part in a systematic survey of the country's soil and water resources, and in efforts to restore productivity to the erosion-ravaged land.

Later, while helping to establish the first settlements in the highlands of the Negev Desert, I had the unique opportunity to witness the compression of four millennia in the history of land and water management into a mere score of years. I lived for a while with the Bedouin, who at that time were wandering bands eking out an austere existence by grazing emaciated goats and camels on the sparse shrubs of the rock-strewn slopes. Their mode of life resembled that of the Biblical Patriarchs, and of the 12 tribes of desperate desert nomads who were led by Moses and inspired by the vision of a Promised Land with flowing brooks and lush meadows. To me, developing the Negev's agriculture seemed like a re-enactment of the process by which the ancient Israelites metamorphosed from roaming shepherds into permanent planters, and learned the ways of soil and water-husbandry.[1]

One of my most inspiring early experiences was the chance to associate with a man of great vision who combined idealism with realism in an unique way. On a chance visit to our little village in the desert, David Ben Gurion, Israel's Founding Father and first Prime Minister, made a sudden decision—astonishing for a political leader at the peak of his power—to resign from the government and come to join and work with us at our arduous task of land reclamation.[2] Ben Gurion lived out his life in that desert village, which we called Sdeh-Boker ("Herdsman's Field"), and is buried there. And it was at Sdeh-Boker that he enunciated his credo: "The energy contained in nature—in the earth and its waters, in the atom, in sunshine—will not avail us if we fail to activate the most precious vital energy: the moral-spiritual energy inherent in man; in the inner recesses of his being; in his mysterious, uncompromising, unfathomable, and divinely inspired soul."

A friendship that formed in the mid-1950s between Ben Gurion and U Nu, then Prime Minister of Burma, led to an agreement between the two countries to develop mechanized crop production in the Burmese upland regions.[3] And so it was, following my experience in the Negev, that I was asked to undertake a very different kind of mission to the tropical rain forests and drenched river valleys of Southeast Asia. The task in Burma and later in Thailand was to help initiate permanent cultivation of upland soils in place of the "primitive" local practice of "shifting cultivation." My experience there and later in other parts of the Third World led me to realize the fallacy of our initial, simplistic assumption that, given enough machinery, fuel, chemicals, and know-how from the outside, underdeveloped lands could be reclaimed straightaway and cultivated without any serious environmental, social, and economic problems.

Returning to Israel, I served as head of the Soil Technology Division of the state's Agricultural Research Organization, and later as head of the Soil and Water Sciences Department at Hebrew University. During the 1960s and early 1970s, I took

part in the intensive effort to improve the efficiency of water-
use that resulted in doubling crop yields while reducing average
crop water requirements by one-third—a singular achievement
of the State of Israel.

In the course of my subsequent career with various national
and international agencies, I have had occasion to observe and
experience the management and mismanagement of land and
water on every continent and in varied locations and circumstances, humid and arid, tropical and temperate, in developing
as well as in developed countries. Among my tasks have been
research, consulting, and managerial assignments[4] in such varied
countries as Iran, India, Pakistan, the Philippines, Japan, and
China in Asia; Cyprus, Italy, Belgium, France, and Holland
in Europe; Nigeria, Egypt, the Sudan, Ethiopia, the Ivory Coast,
and Mauritius in Africa; the continent of Australia; as well as
Colombia, Mexico, and Canada in the Western Hemisphere.
In addition, I have worked extensively in the United States on
problems of soil and water management and the protection of
the environment. I have also had occasion to conduct professional
visits to the Soviet Union and the countries of Eastern Europe,
as well as to several countries in sub-Saharan Africa and Latin
America. Throughout these ventures, I have striven to gain a
global perspective regarding the two major dilemmas besetting
the world in our time: the widespread recurrence of famine
affecting many nations, and the evident deterioration of the
environment affecting all nations.

I believe that any rational control over the impact that human
activity has on the environment must be based on a fundamental
understanding of the processes at work. Obviously, we cannot
protect what we do not understand. It is in the interest of
promoting and disseminating such an understanding that I have
undertaken this book. My treatment of the subject, however,
is not meant to be encyclopedic, but illustrative. In selecting
the aspects to highlight, as well as in setting the style, order,
and tone of the exposition, I have had to make personal judgments.
Being an environmental scientist, I have chosen to devote primary
attention to the natural resources of land and water and to the

physical-biological processes (both natural and anthropogenic) governing them. At the same time, I am aware of the importance of social and economic factors that impel human societies to treat these resources as they do.

In writing this book, I have attempted to explain rather than advocate. However, I cannot and should not hide my point of view. I am a committed naturalist who nevertheless supports development, provided it is judicious and environmentally sound. I have tried to strike a balance between the lamenting or castigating tone now fashionable with some environmentalists, and the complacent view of those inveterate optimists who trust blindly that our problems will somehow be solved by a scientific or technological *deus ex machina*. I do not wish to preach an angry jeremiad of fire and brimstone, but to explore the lessons of our past and present relation to the earth from whence we came, and to which we must return—in spirit as well as in body.

*Then God Yahweh formed man out of
the soil of the earth*

GENESIS 2:7

2

MAN'S ROLE ON GOD'S
EARTH

WE LIVE ON a unique planet bathed in the light and warmth
of a nearby star we call the sun. Alone among the planets
revolving around that star, ours is endowed with the
fortuitous—though ever tenuous—combination of conditions capable of generating and sustaining the miracle of life. And what
a rich and abounding variety of life our earth has spawned! It
includes millions of types of creatures, each unique in form
and function, yet all engaged interdependently in an elaborate
dynamic performance, like players in an enormous philharmonic
orchestra. Altogether, the multitude of plants and animals coexist
both competitively and cooperatively in a more or less stable
community self-regulated by an intricate set of checks and balances.

Pondering the intrinsic mutuality of life on earth, one cannot
but wonder at the discordant anomaly that has so recently intruded
upon nature's pluralistic harmony: How did one species gain
such overwhelming dominance over so many others, indeed over
the very processes that control all life? And how could the members of this clever species fail so utterly and for so long to
realize the dire consequences of their carelessly exercised dominance?

11

The Hebrew Bible provides a profoundly symbolic account of the act of creation, the beginning of life on earth and the origin and role of humankind. It describes how, after summoning up radiant energy ("Let there be light!"), the Creator imposed form and order upon the primeval chaos by separating land from water, and earth from sky. The sea and the land were then made to generate a myriad living species, and man—the presumed pinnacle of creation—was granted a privileged status in the hierarchy of life. This much of the account is known by all.

Less widely noticed is the curious fact that the first two chapters in the Book of Genesis actually give not one but two accounts of creation. Of many contradictions between the two,[1] for us the most significant is the role assigned to humans in the scheme of life on earth. In the first chapter we read that God (called by the plural name "Elohim") decided to "make man in our own image, after our likeness, and let them rule over the fish of the sea and over the fowl of the air, and over the cattle, and over all the earth, and over every creeping thing that creepeth upon the earth." And God blessed man and woman and said unto them: "Be fruitful, and multiply, and *fill the earth, and conquer it;* and rule over the fish of the sea and the fowl of the air and every animal creeping over the earth." And furthermore God said: "Here, I have given you every herb yielding seed and every tree with fruit . . . to you it shall be for food." All this can be construed as a divine ordination of humans to dominate the earth and use everything on it for their own purpose.

But the act of creation and the divine injunction to man are described quite differently in the second chapter of Genesis: "God Yahweh[2] formed man out of the soil of the earth and blew into his nostrils the breath of life, *and man became a living soul.*[3] And God Yahweh planted a garden in Eden in the east and placed the man therein." Then comes the crucial statement: "God Yahweh took the man and put him in the Garden of Eden *to serve and preserve it.*"[4] Here, man is not given license to rule over the environment and use it for his purposes alone, but—quite the contrary—is charged with the responsibility to nurture and protect God's creation.

Thus, latent in one of the main founts of Western Civilization we have two opposite perceptions of man's destiny. One is anthropocentric: man is not part of nature but set above it. His manifest destiny is to be an omnipotent master over nature, which from the outset was created for his gratification. He is endowed with the power and the right to dominate all other creatures, toward whom he has no obligations. In the words of the 115th Psalm: "The heavens are the Lord's, but the earth He hath given to the children of man." The same notion was expressed by Protagoras: "man is the measure of all things."

The other view is more earthly and modest. Man is made of soil and is given a "living soul," but no mention is made of his being "in the image of God." Man is not set above nature. Moreover, his power is constrained by duty and responsibility. Man's appointment is not an ordination but an *assignment*. The earth is not his property; he is neither its owner nor its master. Rather, man is a custodian, entrusted with the stewardship of God's garden, and he can enjoy it only on the condition that he discharge his duty faithfully. This view of humanity's role accords with the modern ecological principle that the life of every species is rooted not in separateness from nature but in integration with it.

Alas, as the story unfolds, Adam and Eve soon abused God's trust, succumbing to short-sighted temptation and consuming beyond their needs, rather than preserving the Garden of Delight. Perhaps in so doing they banished themselves from God's garden by despoiling it, so that it was no longer Eden but indeed a "cursed land." They and their descendants were thenceforth condemned to suffer the consequences. They were sentenced to a life of toil: "Cursed is the earth for thy sake. . . . With the sweat of thy brow shalt thou eat bread till thou return to the earth, for out of it wast thou taken, for soil thou art and unto soil shalt thou return."

Over the generations, it has generally been the arrogant and narcissistic view, implied in the first Biblical account, that has prevailed. It has repeatedly been cited and used as a religious justification or rationale for man's unbridled and relentless exploitation of the environment.[5] The question now is whether we have learned our lesson and are ready at last to accept the long-ignored second view of our proper role in relation to nature.

Readers of the Bible in translation miss much of the imagery and poetry of the evocative verbal associations in the original Hebrew. The indissoluble link between man and soil is manifest in the very name "Adam," derived from *adama*—a Hebrew noun of feminine gender meaning earth, or soil.[6] Adam's name encapsulates man's origin and destiny: his existence and livelihood derive from the soil, to which he is tethered throughout his life and to which he is fated to return at the end of his days. Likewise, the name of Adam's mate, "Hava" (rendered "Eve" in translation) literally means "living." In the words of the Bible: "And the man called his wife Eve because she was the mother of all living." Together, therefore, Adam and Eve signify "Soil and Life."

The ancient Hebrew association of man with soil is echoed in the Latin name for man, *homo*, derived from *humus*, the stuff of life in the soil. This powerful metaphor suggests an early realization of a profound truth that humanity has since disregarded to its own detriment. Since the words "humility" and "humble" also derive from *humus*, it is rather ironic that we should have assigned our species so arrogant a name as *Homo sapiens sapiens* ("wise wise man"). It occurs to me, as I ponder our past and future relation to the earth, that we might consider changing our name to a more modest *Homo sapiens curans*, with the word *curans* denoting caring or caretaking, as in "curator." ("Teach us to care" was T. S. Eliot's poetic plea.) Of course, we must work to deserve the new name, even as we have not deserved the old one.

Other ancient cultures evoke powerful associations similar to those of the Hebrew Bible. In the teachings of Buddha, not only the earth itself but indeed all its life forms (even those that may seem lowliest) are spiritually sacred. To the ancient Greeks, the earth was Gaea, the great maternal goddess who, impregnated by her son and consort Uranus (god of the sky), became mother of the Titans and progenitor of all the many gods of the Greek pantheon. Among her descendants was Demeter, the goddess of agriculture, fertility, and marriage. The story

of Demeter's daughter, Persephone, and—in a different context—
the Egyptian god Osiris, symbolized the annual cycle of death
and rebirth. In the ancient nature cults the major deity of the
earth was generally feminine.[7] The earth was seen as the source
of fertility, the site of germination and regeneration, indeed
the womb of life. And when the plow was invented, its use to
penetrate the soil and open it for seeding seemed to simulate
the very act of procreation.

Worship of the earth long predated agriculture and continued
after its advent. The earth was held sacred as the embodiment
of a great spirit, the creative power of the universe, manifest
in all phenomena of nature. The earth spirit was believed to
give shape to the features of the landscape and to regulate the
seasons, the cycles of fertility, and the lives of animals and
humans. Rocks, trees, mountains, springs, and caves were recognized as receptacles for this spirit, which the Romans attributed
to their earth goddess, Tellus.

The cult of the earth spirit is perhaps the oldest and most
universal element in all religions. The Australian aborigines
and the African Bushmen, among the last to have maintained
the pre-agricultural hunter-gatherer mode of life, have always
sanctified and revered the earth as the great provider, the source
of all inspiration and sustenance. So did the American Indians.
In 1852, when the United States Government wished to purchase
the land of the Indian tribes in the Northwest, their Chief
Seattle sent back this eloquent reply[8]:

How can we buy or sell the sky or the land? The idea is strange
to us. If we do not own the freshness of the air and the sparkle
of the water, how can you buy them? Every part of this earth is
sacred to my people, every shining pine needle, every sandy shore,
every mist in the dark woods, every meadow, every hum-
ming insect. . . . We are part of the earth and it is part of
us. . . . What befalls the earth befalls all the sons of the earth.
This we know: the earth does not belong to man, man be-
longs to the earth. Man did not weave the web of life, he is
merely a strand in it. To harm the earth is to heap contempt
upon its creator.

Seattle's words reverberated in the dignified protest by the American Indian prophet, Smohalla (circa 1872), against the proposal to turn his people (the Plateau Indians of the Northwest, who had been hunters) into cultivators: "You ask me to plow the ground. Shall I take a knife and tear my mother's breast?"[9]

Other cultures and religions did not consider agriculture to be a violation of the earth, but—quite the contrary—a way to make the earth happy and fruitful. Such, for example, was the prevalent belief in ancient Persia. The belief that agriculture is necessarily good, however, ultimately became self-defeating. The hillsides of Persia, like those of other uplands in the Near East and around the Mediterranean, were deforested and subjected to erosion, while the irrigated bottomlands, like those of Mesopotamia, suffered silting salinization.

As soil is the material substrate of life, water is literally its essence. In a symbolic sense, water *is* life. A spring of water bubbling up from the ground seems indeed to be alive, and one can easily perceive why it has always inspired animistic and divine associations. It was "living water" to the Hebrews, "running water" to the Arabs. "And with water we have made all living things," states the Koran. The Egyptian priests posited that the earth itself was created out of the primordial waters of Nun, and that such waters still lay everywhere below the soil. Noticing that the delta was being augmented by the Nile, the Egyptians could easily come to believe that their land was being produced by the river's water, transmuted into solid earth. A similar notion was prevalent among the Mesopotamians. According to a Babylonian legend, all the world was originally sea, until Marduk bound a rush mat upon the face of the waters and piled soil on it. This is reminiscent of the Hebrew Bible's depiction of the initial state of chaos, when "the earth was unformed and void, and darkness was upon the face of the deep, and the spirit of God hovered over the face of the waters." The Greeks also believed that the earth continued to be surrounded by an endless expanse of primeval waters extending beyond the sea.

To signify water, the ancient Egyptians drew wavy lines

reminiscent of wriggling snakes. The symbol is preserved in
the Hebrew and Phoenician letter "mem" (representing *mayim*,
or water), which in turn became the Latin letter M.[10] The
Sumerian word for water was *a*, which also signifies sperm, or
generative power: the masculine element that fructifies earth.
Among the rivers and springs that are held sacred in many
countries, most notable are the Nile, the Ganges, and the Jordan.
In the Judeo-Christian tradition, as well as in many of the Eastern
religions, water is regarded not only as a physical cleansing
agent but also as a source of spiritual purification and renewal.

Our interest in how soil and water function in the biosphere,
and in how they can be managed or mismanaged, derives as
much from necessity as from innate scientific curiosity. Throughout the history of civilization, the pressure of increasing population has led to the careless exploitation of the world's most
valuable soil and water resources, and at times to their rapid
destruction. Superficial observers of history who ignore the role
of environmental factors may ascribe the defeat of an empire to
moral decay, cultural enfeeblement, lead poisoning, or lack of
military preparedness—when actually the main contest had already been decided by the abuse and degradation of vital resources.

The failure to heed the lessons of the past is reflected in
the Koran: "Do they not travel through the earth and see what
was the end of those before them? . . . They tilled the soil
and populated it in great numbers . . . there came to them
their apostles with clear signs, which they rejected, to their
own destruction. It was not Allah who wronged them, but
they wronged their own selves" (Sura XXX: 9).

Today there is clear and urgent reason for us to be concerned
over the adequacy of land and water resources to satisfy the
demands of our own profligate civilization. Our concern is not
merely for the availability of these resources but for their quality
as well. The encroachment of urban, industrial, transportation,
and even recreational activities on the landscape, along with
the application of "efficient" modern techniques of agriculture,
construction, mining, and waste disposal, exert growing pressure
on the limited resources of good land and water.

Among the many nations abusing their natural endowment, America is not the least offender. This country's fundamental strength depends on its great soil and water resources, and their wasteful and destructive exploitation is surely sapping the nation's innate strength and jeopardizing its future.

We can take no comfort at all in the fact that the problem is universal. Absurdly, nations fight wars over every inch of their political boundaries while mindlessly sacrificing whole regions to environmental degradation. Their patriots salute the flag and take up arms to defend their country against external enemies, while neglecting its environment and ignoring the real attacks being waged from within on the land they purport to love. Thousands of years are required for a soil to form in place, yet this amazingly intricate work of nature can be destroyed by man, with remarkable dispatch, in just a few decades. We must understand that, on the time-scale of human life, the soil is a non-renewable resource. So is a mature forest, a river, a lake, or an aquifer. They belong not only to those who are the titled owners at this moment, but to future generations as well. In an even more profound sense, both soil and water belong to the biosphere, to the order of nature, and—as one species among many, as one generation among many yet to come—we have no right to destroy them.

There is, however, a special fascination in the topic of soil and water that transcends both our scientific interest and our concern over contemporary problems. This fascination underlies every child's instinctive interest in mud pies. It might even antecede our early ancestors' discovery that mud can be used to construct homes and temples, and can be used as well to form a new and lasting material—the first man-made substance, called "ceramic." There is an even greater meaning to the soil. In some intuitively perceivable sense, the quest for a deeper understanding of the soil's role in the natural environment and in the life of humanity is more than an intellectual exercise or a merely utilitarian task. It might even be something of a spiritual pilgrimage, impelled by an ancient call, a yearning to return to a life of greater authenticity. In Homer's words: "I would

rather be tied to the soil as a serf . . . than be king of all
these the dead and destroyed."[11]

Can greater awareness of our environment and of our place
in it help awaken us from our narcissistic indulgence, and foster
a more appropriate sense of humility toward nature? And can
this sense bring us any closer to our common physical, biological,
and cultural moorings? Can it reconnect us spiritually with our
humble origins, from which we have for so long been separated
yet never completely severed?

An awareness that human activities are now destabilizing
the biological balance on a global scale does not require us to
romanticize the primitive life, as did Jean-Jacques Rousseau,
and to advocate a total renunciation of materialism and progress.
There probably never was a golden age of perfect human happiness
in a pristine world, which in any case could not have lasted in
the face of the innate aptitude of humans for meddling with
the environment, and their urge to enlarge their domain and
multiply in number. Nor, on the other hand, need we accept
the notion expressed by Thomas Hobbes that modern man is
infinitely better off than the "primitive" child of nature whose
life was "solitary, poor, nasty, brutish, and short." These dia-
metrically opposed notions are both equally simplistic. Clearly,
however, something has gone wrong in our relation with nature,
and it behooves us to ponder what it is and how it started.

We live in an age and culture that is very sensitive to human
rights, but does not grant equal weight to human responsibilities.
We insist on our prerogatives, and neglect our obligations. Our
attitude toward the environment is marked by careless confidence
and reckless self-indulgence. These are attitudes and actions that,
in any individual, we recognize as childish. And just as a mature
person must learn to consider the circumstances and needs of
others, so a mature society must restrain its exploitation of resources and consider both the rights of future generations and
the needs of other species.

Perceiving that modern civilization had drifted from that
intimacy with the natural elements that was the mark of our
forebears, Friedrich Nietzsche felt driven to proclaim: "Man

and man's earth are unexhausted and undiscovered. Wake and listen! Verily, the earth shall yet be a source of recovery. Remain faithful to the earth, with the power of your virtue. Let your gift-giving love and your knowledge serve the meaning of the earth." Perhaps our most precious and vital resource, both physical and spiritual, is that most common matter underfoot which we scarcely even notice and sometimes call "dirt," but which is in fact the mother-lode of terrestrial life and the purifying medium wherein wastes are decomposed and recycled, and productivity is regenerated.

A glimpse of earth from space should be sufficient to restore the true perspective. It shows the planet whole, without political or tribal boundaries. How beautiful, how colorful, how delicate is this ball of lapping waters, floating continents, and swirling clouds gliding in a thin veil of air. And how small, unique, and solitary is this one and only home of ours. We must listen to its signals of distress, for it is our parent and we are all its dependent children.

PART II

THE

NATURE

OF SOIL

AND

WATER

To see a world in a grain of sand
and a heaven in a wild flower

WILLIAM BLAKE, *Auguries of Innocence*[1]

3

THE FERTILE
SUBSTRATE

A ROMANTIC POET gazing through his window at a green field outside might view it as a place of idyllic serenity. Not so the environmental scientist, who discerns not rest but unceasing turmoil, a seething foundry in which matter and energy are in constant flux. Radiant energy from the sun streams into the field, and as it cascades through the atmosphere-plant-soil continuum it generates a complex sequence of processes: Heat is exchanged; water percolates through the intricate passages of the soil; plant roots suck up some of that water and transmit it through the stems to the leaves, which transpire it back to the atmosphere. The leaves also absorb carbon dioxide and synthesize it with soil-derived water to form the primary compounds of life. Oxygen emitted by those leaves makes the air breathable for animals, which consume and in turn fertilize the plants. Organisms in the soil recycle the residues of both plants and animals, thus releasing nutrients for the renewal of life.

The crucible of this foundry is the soil, a rich mix of mineral particles, organic matter, gases, and nutrients which, when infused with vital water, constitutes a fertile substrate for the initiation and maintenance of life. The soil is thus a self-regulating

23

biological factory, utilizing its own materials, water, and energy from the sun. It also determines whether water from rainfall and snowfall reaching the earth's surface will flow over the land as runoff, causing flash floods, or seep downward toward the subterranean reservoir called groundwater, which in turn maintains the steady flow of springs and streams. The soil, with its capacity to store moisture, buffers and moderates these phenomena, while serving as a bank for the water requirements of growing plants.

The soil also acts as our earth's primary cleansing and recycling medium, in effect as a "living filter," wherein pathogens and toxins that might otherwise foul our environment are rendered harmless and transmuted into nutrients. Since time immemorial humans and other animals have been dying of all manner of diseases and then been buried in the soil, yet not a single major disease is transmitted by it. The term *antibiotic* was coined by the soil scientists who discovered streptomycin.[2] Ion exchange, a useful process of water purification, was also discovered by soil scientists studying the passage of solutes through beds of clay.

But just what do we mean by "soil"? A precise definition is elusive, for what we commonly call soil is anything but a homogeneous entity.[3] It is in fact an exceedingly variable body with a wide range of attributes. Perhaps the best we can do at the outset is to define soil as the fragmented outer layer of the earth's terrestrial surface in which the living roots of plants can obtain anchorage and sustenance, alongside a thriving biotic community of microscopic and macroscopic organisms.

Initially, soil is formed through the physical disintegration and chemical decomposition of rocks exposed to the action of the weather—the process called "weathering." But just how does a strong and massive bedrock yield to weathering? It succumbs to a slow but relentless attack by a combination of natural forces, including the stresses that result as rock surfaces repeatedly expand and contract as they experience fluctuations of temperature. Additional stresses are due to the wetting and drying of porous rocks, and the freezing and thawing of water in cracks.

Growing roots penetrate such cracks and exert powerful swelling
forces that further split the rocks. Abrasive particles carried in
creeping ice, flowing water, and blowing wind scour and grind
rock surfaces.

Superimposed upon these physical processes and augmenting
them are numerous chemical reactions that tend to modify still
more the primary mineral material present in the original
bedrock.[4] The chemical names for these processes are hydration,
oxidation and reduction, dissolution and dissociation into ions,
and acidification and alkalinization. The selective removal of
various components occurs through precipitation, volatilization,
and leaching. Some primary minerals present in the bedrock
are relatively resistant to chemical change and may retain their
original mineralogical character even in fragmented form. An
example is quartz, a major constituent of granite that is transformed into sand when that rock disintegrates. Other minerals,
like feldspars or micas, are more reactive, so they are gradually
decomposed and reconstituted into a series of secondary minerals,
formed within the soil.

Among the interesting and important minerals formed in
the soil is a very distinctive group called "clays."[5] The clays
generally occur as microcrystals or as amorphous gels that exhibit
an enormous surface area per unit of mass—as much as 80 hectares
(200 acres!) per kilogram of clay. This great exposure of clay
surfaces within the soil is not inert, but highly reactive, as
each microscopic particle is endowed with an electrostatic charge
analogous to that of an electronic condenser (or capacitor). Hence
the clay is the active fraction of the soil and it takes part in a
series of complex physicochemical and biological interactions.
So pronounced is the physicochemical activity of clay, that some
scientists have raised the intriguing possibility that it might
have played a role in the origin of life on earth. The unique
properties of clay may have enabled it to act as a template for
living molecules, so that the first organisms on earth may have
used clay minerals as forerunners of proteins, nucleic acids, or
other biomaterials.

Clay particles attract positively charged ions, called cations, and repel negatively charged ions, called anions. Important among the ions is the hydrogen ion, which imparts acidity to the soil. Countering soil acidity are the alkaline cations of sodium and potassium, or the cations of calcium and magnesium. The electrostatic attraction of these ions to clay surfaces is an important mechanism in the retention of nutrients in the soil. Clay particles also have a great affinity for water, and as each tiny particle is enveloped with water the entire assemblage tends to swell. Conversely, when water is extracted the clayey soil tends to shrink. This accounts for the troublesome but familiar effect expansive clay has on buildings and roads that rest on it. In some cases, entire cities and regions may suffer from the process of clay-bed subsidence. Within the soil, however, the expansion and contraction of clay produces microfissures that facilitate the formation of aggregates of various sizes, shapes, and orientations.

In comparison with clay, the larger particles in the soil, called "sand," and the intermediate-size particles called "silt," are relatively inert: they do not swell or shrink, nor do they attract or exchange ions. Their presence makes the soil more permeable and easy to till, less sticky or muddy when wet, and less cloddy or tightly cemented when dry. Coarse-grained soils also retain less water and ions. Because they are relatively inert, the sand and silt fractions of the soil can be thought of as the soil's "skeleton," whereas the clay, by analogy, is likened to the "flesh." Together, the entire assemblage of particles, variously associated and spatially organized, constitutes the soil's matrix.

The soil material may accumulate in the place where it was formed, giving rise to a "residual soil;" or it may be picked up and transported by water, glaciers, or wind, to be deposited at some distance from its source. If transported by water, the deposit is called an "alluvial" soil, and if by wind it is called an "eolian" soil. The former is typical of river valleys, the latter of plains adjacent to arid or semiarid regions.

Still, the deposit of fragmented material is not yet a true soil, just as a collection of bricks of various sizes and shapes thrown haphazardly into a heap does not make a building. The

same bricks, differently arranged and mutually bonded, can form
a home or a factory. Similarly, soil material can be merely an
unstable assemblage of random particles, or it can consist of a
distinctly structured pattern of internally bonded aggregates having regular sizes and shapes. The processes of weathering, transport, and deposition of sediments is only a prelude; the series
of biophysical and biochemical processes that follow ultimately
reorganize the initially haphazard mineral soil constituents into
a distinct, internally ordered, living natural body. This final
metamorphosis is brought about by the activity and the accumulated organic products of myriad microscopic and macroscopic
plants and animals, which coinhabit the soil as an integrated
community. We can visualize the living soil as a composite
body in the same way we think of the human body as a distinct
organism, even though in reality it is an ensemble of numerous
interdependent and symbiotically coordinated groups of compounds, cells, organs, and colonies of varied microorganisms.[6]

The gradual disintegration of rocks into particles, and the
further metamorphosis of this loose assemblage into a more or
less stable, functioning soil body, typically requires many hundreds or thousands of years. In this genesis, the particle clusters
that are clumped together by clay are stabilized internally by
glue-like organic gels. The soil body as a whole acquires a characteristic profile, consisting of a sequence of horizons, like a layer
cake. These horizons are formed within the soil, and are therefore
called "pedogenic." As such, they differ from the depositional
layers that are characteristic of recently laid sediments. The upper-
most horizon, or topsoil, termed the A-horizon, is the zone of
major biological activity, and therefore is typically enriched with
decomposed organic matter (humus) and nutrients, so it is generally more fertile than the underlying soil. Here, living plants
and animals and their residues interact with an incredibly diverse
and labile multitude of microorganisms of every kind—such as
bacteria (including the filamenous or rod-shaped types known
as actinomycetes), protozoa, and fungi—billions of which can
be found in a mere handful of topsoil.

In the natural state, this top horizon, which is often discernibly darker than the deeper horizons because of its humus content, may have a thickness of 20 to 40 centimeters (about a foot). When stripped of vegetative cover, however, and pulverized or compacted excessively by cultivation and traffic, this horizon becomes vulnerable to accelerated erosion by water and wind, and may thereby lose half or more of its original thickness. Thus, a fertile soil layer that takes nature many centuries to establish in place can be destroyed by careless human intervention within just a few years. This insidious process of accelerated and practically irreversible erosion is a consequence of removing the layer's natural protection, consisting of plants and their residues, and of tilling the soil excessively. Unless erosion is controlled by stringent soil conservation, it can rapidly degrade the productivity of an initially thriving region. Unchecked erosion may even doom the agricultural industry of an entire nation, thus undermining the foundation of its economic and political strength.

In a mature soil, the A-horizon (topsoil) is underlaid by a subsoil zone called the B-horizon, where some of the mobile materials washed down from the topsoil tend to accumulate. These materials include very fine particles of clay that can be carried in suspension by percolating water, and—in the case of semiarid regions—certain salts of moderate solubility like lime. The B-horizon, generally thicker than the A-horizon, in turn overlies the weathered bedrock, which is the soil's parent material. The depth of the entire soil profile, though it varies, is generally not much more than one meter deep. So incredibly thin, and vulnerable, is this outer layer of the earth's crust that supports all terrestrial life.

This sketchy account of a typical soil profile is admittedly simplistic. In reality, soils formed in different climates on different landscapes of different parent materials vary widely in their characteristics: some are deep and others shallow; some clayey and others sandy; some black and others red, or brown, grey, or yellow; some acidic and others alkaline; some friable and others hard; some highly permeable and others nearly impervious; some highly fertile and others practically sterile. These differences are not due to mere chance, but relate to definable natural factors.

As far as we know, a Russian scientist named V. V. Dokuchaev was the first to recognize, more than a century ago, that soils are distinct natural bodies with regular horizons and characteristic geographic distributions. Independently, though somewhat later, an American scientist named E. W. Hilgard made the same discovery. The work of these pioneers, and of those who followed them, has culminated in a global classification and characterization of soils, based on recognition of the major factors that influence their formation and geographical distribution. The recognizably different types of soil are legion, reflecting the enormous heterogeneity of such determining factors as bedrock composition, landscape (slope), climate, vegetation, and the length of time the soil-forming processes have been at work. Specialists who call themselves pedologists are fond of endlessly reclassifying soils into more and more types, to which they apply strange-sounding names.[7]

Undoubtedly the most important factor influencing soil formation is the climate. Each climatic zone exhibits its own characteristic group of soils. In the humid tropics, for example, there is a tendency to leach away the silica and to accumulate iron and aluminum oxides. As a result, the soils are typically colored red, the hue of iron oxide. Chunks or blocks excavated from such soils and dried in the sun may harden to form bricks; hence these soils are called *laterites,* from the Latin word *later,* meaning brick. On the other hand, soils of humid cool regions often exhibit an A-horizon consisting of a thin surface layer darkened by organic matter and underlaid by a bleached, ash-like layer; in turn, this overlies a clay-enriched B-horizon. These soils are called *podzols,* from the Russian words *pod* (ground) and *zola* (ash).

In contrast with the soils of humid regions, from which nearly all readily soluble salts have been leached, the soils of arid regions tend to accumulate the moderately soluble salts of calcium and magnesium, and sometimes even the more readily soluble salts of sodium and potassium. While such soils may retain more nutrients and thus are often more fertile, they are also prone to excessive salinity. An outstanding soil found in some intermediate semihumid to semiarid regions, such as the Ukraine in the USSR and the prairie in the United States and Canada, is the so-called *chernozem* (Russian for "black earth"), with its unusually thick and nutrient-rich layer of organic matter. Chernozem is one of the most fertile soils in the world: deep, stable, and easily tilled. It is the soil of the cornbelt and the wheat belt in the United States, and the area of primary grain production in the Soviet Union.

The earth's land surface is a variegated tapestry spanning a wide diversity of soils. Any rational attempt to manage such differing and labile natural bodies efficiently and sustainably must be tailored to their specific characteristics and internal workings. These workings are intimately associated with the presence and movement of water, a substance indispensable to life.

And with water we have made all living things.

THE KORAN, SURA XXI (ANBIYAA):30

4

THE VITAL FLUID

LIQUID WATER is the most plentiful substance on earth, covering more than two-thirds of its surface in oceans, seas, and lakes. Even the continental areas are frequently charged with and shaped by water. In vapor form, water is always present in the atmosphere. It is our planet's most distinctive and active agent.

Although the importance of water was recognized early in the history of civilization, and our ancient forebears revered it, not much was known about its real nature.[1] In the Middle Ages people believed that fresh water emanated magically from the bowels of the earth. They could not imagine that all the water flowing in innumerable springs and mighty rivers could possibly result from so seemingly feeble a source as rain and snow. The first to conjecture this was Leonardo da Vinci, but it was only in the latter part of the seventeenth century that the English astronomer Edmond Halley and, separately, the Frenchman Claude Perrault, proved the principle by measurement and calculation. Water was long thought to be a single element, until early in the eighteenth century, when it was found to consist of a combination of hydrogen and oxygen atoms.[2]

31

Life as we know it began in an aquatic medium, and water is still the principal constituent of all living organisms. "Life," wrote Vladimir Vernadsky some 100 years ago, "is animated water."[3] Though we appear to be solid, we are in fact liquid bodies, similar in a way to gelatin, which also seems to be solid but is in fact largely water, "gelled" by the presence of an organic material. The analogous material in our bodies is protoplasm, consisting of various amino acids and proteins. Many components of living bodies are literally dilute suspensions of these vital organic compounds in water. Actively growing herbaceous plants are mostly water. A human infant born into the world is 90 percent water in volume, and much more than that in terms of the proportional number of molecules. And although as we grow older we tend to dry up a bit, we are still mostly water—nearly 40 liters of it (about 10 gallons), encapsulated in trillions of cells.

Water and its ionization products, hydrogen and hydroxyl ions, are important factors determining the structure and biological properties of proteins and other cell and tissue components. The basic processes of life are intrinsically dependent on water's unique attributes. Water flows in the veins and roots of all organisms, the lifeblood of their collective body. Far from being a bland, inert liquid, water is a highly reactive substance and an exceedingly effective solvent and transporter of numerous substances, which it conveys in solution or in suspension.

Water, however, is still something of an enigma, possessing unusual and anomalous attributes not entirely understood. Perhaps the first anomaly is that water, despite its low molecular weight, is a liquid and not a gas at ordinary temperatures. (Its sister compound, hydrogen sulfide, boils at $-60.7°$ Celsius.) Compared with other common liquids, water has many unusual physical properties, including high melting and boiling points, heats of fusion and vaporization, specific heat capacity, dielectric constant, viscosity, and surface tension. In particular, the specific heat capacity, which is the amount of heat required to raise the temperature of a unit mass by one degree, is 10 times as high in water as in iron, 30 times as high as in mercury, and

5 times as high as in dry soil. These anomalous properties suggest the existence of an especially strong force of attraction between the molecules in water that imparts such a high internal cohesion to this liquid.

The strong intermolecular force in liquid water is caused by the electrical polarity of the water molecule. The H-O-H link within each molecule is not linear but bent at an angle of about 105°, so the molecule is dipolar, like a tiny magnet with one side electrostatically more positive and the other side more negative. This is the secret of the intermolecular affinity: Every hydrogen nucleus, while it is attached by marriage to its own molecule, is also attracted to the oxygen of a neighboring molecule, with which it forms a casual secondary association called a "hydrogen bond." However, unlike the chemical bonds of solids, which join molecules in a crystalline lattice that is fixed in space, those bonds linking water molecules to each other shift position repeatedly. A molecule changes partners 10 billion to 100 billion times a second, thus forming and dissolving "flickering clusters" continuously.

One noteworthy property of water is that it is more dense as a liquid than as a solid. Also unusual is that it has its greatest density just a few degrees above its freezing point. These anomalies have important consequences. For instance, when water freezes, its solid form (ice) rises to the surface rather than sinking to the bottom, as would most freezing liquids. This has the effect of insulating the surface of lakes, so that the bottom remains liquid and permits aquatic life to survive through winter. The expansion of water on freezing helps to disintegrate rocks, but also poses a danger to living cells, which might rupture if subjected to freezing. Water in living cells is, however, less apt to freeze than the water in its "free" state in streams and lakes. It is modified by its proximity to cell surfaces, to which it is attracted, and under whose influence it changes its internal structure and density. Water modified in this manner is called "vicinal," after the Latin *vicinalis,* meaning neighboring.

The unique attributes of water have a strong effect on the climatic environment. Heat stored in the waters of lakes and oceans helps to regulate temperature fluctuations over adjoining land masses. Warm ocean currents convey heat from the tropics to higher latitudes. The whiteness of snow and clouds serves to reflect part of the sunlight back into space, thus moderating the heat balance of the earth. On the other hand, water vapor in the air has the effect of blocking some of the heat emitted from the earth's surface, so it is a natural contributor to the so-called greenhouse effect.[4] The condensation of water vapor to form masses of clouds, the inter-regional transfer of clouds by winds, and the tendency of clouds to precipitate their moisture in the form of rain or snow, all serve to distribute heat and moisture over the earth's surface, and thus have much to do with determining the pattern of climate.

Water is the universal solvent. Numerous substances, both mineral and organic, dissolve readily in water, and its presence promotes many chemical reactions, such as the weathering and decomposition of minerals and rocks. Water transports nutrients in the soil, as well as into and through plants. Water delivers vital nutrients and oxygen in our bodies. It also carries away the wastes of our bodies, towns, and factories. As a solvent, water may contain harmful concentrations of such substances as salts. Hence the importance of considering the quality, as well as the quantity, of water available for human use.

Water has profoundly affected the course of human history. Its abundance has helped societies to flourish, its scarcity has caused them to wither. No consideration of history or of the fate of societies, past or present, can ignore its role. It therefore behooves us to consider the status of water supplies in global terms, for although water covers so much of the earth's surface, it is not universally found in the quality and amount needed. The important facts worth noting are that the total amount of water on earth may seem enormous, but only a small fraction of it is fresh water, and only a fraction of that is accessible to us. As the planet's population grows, and as the world's economy develops, more people make more demands on the same amount of available water. We are also depleting and contaminating some of the surface and subsurface reservoirs on which our future depends.

Since water is a highly mobile and hence an elusive resource capable of moving across property lines and even international boundaries, its management is difficult and its ownership hard to define. As it moves in surface streams or in subterranean strata, it can be accessed by different potential users, who more often compete than cooperate. Each user or group of users can deplete or pollute the water supply available to competitors. Since the essence of a property right is exclusive use, and water is hard to control, there have always been and still are bitter disputes over water rights, more so now than ever. The very word "rival" was originally a Roman legal term referring to a person who shared with another the water of a *rivus,* or irrigation stream.

It is remarkable that certain earlier civilizations, technologically so much less advanced than ours, could thrive in conditions of extreme water shortage—in arid regions and even in deserts. As we will see, however, not all societies have managed water equally well. Our own civilization is now being tested in regard to its management of water as well as soil.

All the rivers run into the sea; yet the
sea is not full.

ECCLESIASTES 1:7

5

THE DYNAMIC CYCLE

THE EARTH AND THE ATMOSPHERE are engaged in an endless reciprocal passing game whose main article of exchange is water. The forces that impel this exchange are the sun's radiant energy and the earth's gravitational pull. Water evaporates from wet ground, from the foliage of growing plants, and from lakes, reservoirs, and seas. It is carried in the air as vapor, which recondenses as a liquid and falls as rain, which in turn feeds the rivers that return water to the ocean. Evaporation from land and ocean sends water back to the atmosphere, and the exchange goes on repeatedly: water circulates from earth to atmosphere to earth. This cyclic exchange is called the hydrologic cycle, and its study is the science of hydrology.

The soil plays a major role in the series of hydrological processes determining the fate of water in the terrestrial environment. The soil surface zone is where the decisive partitioning of rainfall into surface and subsurface water occurs. It is here that the processes of infiltration, runoff, groundwater recharge, evaporation, and soil moisture-extraction and transpiration by plants are generated. In effect, the soil's top layer acts as a sensitive multiple valve, automatically allocating water to these

different functions at time-variable and space-variable rates. One
needs to look into the soil to see how this happens.

The movement of water in the field is a continuous, repetitive
sequence of processes, without beginning or end. For our own
convenience, however, we can visualize the sequence as beginning
with the entry of water into the soil surface zone, continuing
with the temporary storage of water in the soil, and ending
with its removal from the soil by drainage, evaporation, and
plant uptake.

Soil and water have a physical affinity. Dry soil is "thirsty,"
sucking up water the way an old-fashioned blotter sucked up
ink. When the soil surface is wetted by rain, the suction force
of the deeper soil layers, augmented by the force of gravity,
draws the water downward. The soil drinks the rain in a process
called "infiltration." The maximum rate at which the soil is
able to absorb water applied to its surface is called the soil's
infiltrability. It is greatest when the soil is dry, and diminishes
gradually as the soil is wetted to progressively greater depth.
Since the water permeating and seeping in the soil must make
its way through the intricate labyrinthine passages between the
irregularly shaped and oriented soil grains, it is obvious that a
soil's infiltrability depends on the widths and tortuosities of
these interstices, called pores. Hence, coarse-grained soils such
as gravel or sand with wide pores absorb rain more rapidly
than do fine-grained soils such as clays which have very narrow
pores. So do finer-grained soils that have open worm holes,
root channels, and a surface layer (topsoil) with stable aggregates
and wide inter-aggregate cavities—all of which are often eliminated, however, when the soil is manipulated excessively by
tillage and traffic.

Whenever the rainfall rate exceeds the infiltration rate, the
excess rainfall tends to accumulate on the soil surface. If the
surface is level, this excess surface water will remain ponded
until it eventually infiltrates or evaporates after the rain ceases.
If the surface is sloped, however, the excess surface water will
naturally begin to trickle downslope as surface runoff. Over a
smooth surface, runoff takes place initially as uniformly dis-

tributed sheet flow. But as the running water gains volume and momentum, its erosive power increases and it tends to scour the surface, carving out flow paths or channels called rills, which can grow into larger gullies. Numerous gullies converge downslope in a dendritic pattern, combining to form larger and larger rivulets that, in turn, discharge into streams and finally into regional rivers.

As the intensity and duration of a rainstorm increase, the runoff gains in volume and speed and its erosive power multiplies. The running water picks up a load of sediment, and that sediment is carried along by the flood and may be deposited in the lower reaches of a stream. When even a moderate rainstorm breaks out over an entire watershed, runoff from numerous converging tributaries may swell the streams. When a storm is intense, prolonged, and widespread, all tributaries flow simultaneously and a disastrous flood may occur as rivers overflow their banks.

The process of erosion itself begins even before the onset of runoff, under the seemingly benign raindrops that wet the ground and sustain the life of the soil. Each raindrop carries kinetic energy in proportion to the product of its mass and the square of its velocity. Plants and their residues shield the soil by intercepting the rain, thus moderating its force. But if the surface is laid bare by the removal of protective vegetation, it is bombarded directly by myriads of raindrops, each acting as an explosive missile. The rain can then wreak havoc. The soil at the surface slakes down and becomes a layer of slick mud, particularly if it has been loosened by tillage. Aggregates or clods collapse under the onslaught, and clog the surface cavities, thus further reducing the soil's infiltrability and inducing even more runoff. Particles detached and bounced from the surface by the pelting raindrops are splashed downslope and then carried off by the running water. To an onlooker during an intense rainstorm, it might appear as though the soil surface were melting away.[1]

In the wake of a rainstorm, the amount of water a soil can store and make available for plant roots depends in the first place on the amount infiltrated during the storm, and subsequently on the amount lost by evaporation and by internal drainage. The latter is the downward percolation and redistribution of soil water that continues for some time after infiltration has ended. The rate of this internal drainage (rapid at first, but tending to slow down until after a few days it becomes practically imperceptible) determines the effective reserve of soil moisture available to plants during the dry spells between rain events. In arid regions, the supply of rainwater may simply moisten the soil's upper layer, with hardly any excess left to percolate beyond the root zone. In humid regions, however, where the rains are frequent and abundant, much of the infiltrated water will naturally escape internally to depths beyond root penetration, and continue to seep downward toward the groundwater aquifer.[2]

An aquifer (literally, "water carrier") is a geological stratum that can accumulate and transmit water in sufficient amounts to serve as a water source for human use. In other words, it is a saturated layer, occurring at some depth, that can convey water to springs, streams, and wells. Aquifers vary greatly in their characteristics; some consist of unconsolidated coarse sediments such as sand and gravel, and some are pervious or fissured bedrock such as sandstone or limestone. A distinction can be made between confined and unconfined aquifers. A confined aquifer, called "artesian," is one that is overlaid by an impermeable layer.[3] Such an aquifer is generally recharged, not via its impervious ceiling, but from areas some distance away where the permeable formation outcrops. In contrast, an unconfined aquifer is not capped, so the water level is free to fluctuate up and down periodically, depending on the relative rates of charge and discharge.

We can generally gain access to groundwater by digging or drilling a hole through the soil and the underlying strata to the saturated zone. Water then tends to seep into the hole, and the level at which the groundwater comes to rest is called the "water table."[4] The depth of the water table can be highly variable—in places it may be hundreds of meters deep, whereas in other places it may rise to the soil surface. In humid regions, the water table naturally tends to occur at shallower depths than in arid regions. Where the water table is very high, it may invade the rooting zone of the soil and cause it to be waterlogged. The roots of most crops then suffer from restricted aeration. High water table conditions may also occur in arid regions, particularly in river valleys where rapid seepage occurs from river beds as well as from excessive irrigation. In such regions, the groundwater may be brackish and the rise of the water table may induce upward capillary seepage of salt-bearing water to the soil surface, where the water evaporates leaving the salt behind. Thus, groundwater of good quality occurring at an optimal depth can serve as a beneficial source of water, whereas its occurrence at too shallow a depth can pose a threat of soil degradation.

The dynamic interrelationship of soil, water, plants, and humans is described most succinctly in the Koran: "We send down water from the sky according to measure, and we cause it to soak in the soil; and we surely are able to drain it off. With it we grow for you gardens of date-palms and vines; in them ye shall have abundant fruits, and of them ye shall eat" (Sura XXIII: 18–19).

As a tree planted by the waters that
spreads its roots by the stream . . . it
shall not be anxious in a drought, nor
cease from yielding fruit.

JEREMIAH 17:7

6

THE PRIMARY
PRODUCERS

GREEN PLANTS ARE nature's only autotrophs, able to create living matter from inorganic raw materials. Terrestrial plants do this by combining atmospheric carbon dioxide with soil-derived water while converting solar radiation into chemical energy through photosynthesis. This process not only produces food for all the world's heterotrophs,* but also releases into the atmosphere the elemental oxygen animals need to breathe. Water plays a central role in the metabolism of plants, as a constituent of photosynthesis, as a solvent and conveyor of transportable nutrients and enzymes, and as a major structural component often constituting more than 90 percent of the vegetative biomass.[1]

Plants live simultaneously in two very different realms: the bright, rarified atmosphere above; and the dark, dense soil below. In each of these realms, conditions vary constantly, but not necessarily in conjunction. The aerial canopies of plants are designed to reach out so as to intercept and collect sunlight and carbon dioxide, both of which are diffuse rather than concentrated.

---

* Beings, including humans and all other animals, that require complex organic compounds of nitrogen and carbon for metabolic synthesis.

41

To do so effectively, plant canopies tend to maximize their surface of exposure by branching and extensive foliation.

Even more striking than the arms-outstretched stance of the plant canopies is the shape of their root systems, designed like thin tentacles that proliferate and ramify throughout a large volume of soil while exposing an enormous surface area.[2] A single tree can develop a root system with a total length of several hundred kilometers and a total surface area of hundreds of square meters. Such extensive exposure is needed because roots have to suck up water and nutrients from a medium that contains only a meager supply of water per unit volume and only very dilute concentrations of nutrients. And while the atmosphere is a well-stirred and thoroughly mixed fluid, soil moisture is a sluggish and unstirred fluid that moves toward the roots at a grudgingly slow pace, so that the roots have no recourse but to move toward it. Therefore, roots forage constantly for water and nutrients through as large a soil volume as they can reach, so long as the soil is not excessively dry, dense, saline, toxic, or anaerobic.

A seemingly tranquil tree standing in the field is really a highly restless organism in continuous motion: its leaves orienting themselves to intercept sunlight, and fluttering in the breeze while absorbing carbon dioxide and releasing water vapor. At the same time, its roots penetrate ever deeper as they make their way through the maze-like passages of the soil.

Only a small fraction of the water that plants absorb from the soil is used productively in photosynthesis; most, often as much as 99 percent, is lost as vapor in the process known as plant transpiration. Transpiration occurs through minute pores in the epidermis of leaves, called "stomates" (from the Greek word *stoma,* meaning mouth), that are the entry ports for carbon dioxide (CO_2). Transpiration is made inevitable because the same extensive exposure of leaf surfaces that is necessary to facilitate absorption of atmospheric CO_2 and of solar radiation also subjects the moist cells of the leaves to the evaporative power of dry air. From the point of view of crop growers in semiarid and arid regions, transpiration is hence a necessary evil.

Most crop plants are extremely sensitive to any lack of the soil moisture they need to replace the amount lost in transpiration. Water deficits impair plant growth and, if extended in duration, can be fatal. A plant not able to draw water from the soil at a rate that equals the atmosphere's evaporative demands will soon begin to dehydrate and wilt. The problem is especially acute in arid regions, where the relentless demands of an unquenchably thirsty atmosphere can hardly be matched by the sporadic rainfall. To survive during the long dry spells between rains, the plant must rely on the diminishing reserves of moisture retained in the pores of the soil, which itself loses water through direct upward evaporation as well as through downward seepage.

Plants in an arid environment are in the same situation we would be in, if we were living under a government that taxed away 99 percent of our income while requiring us to keep our reserves in a bank that is daily being embezzled. That is why irrigation is so crucial to crop production in arid regions.

Plant roots are also sensitive to excessive water in the soil. Most terrestrial plants are not able to transfer oxygen internally from their above-ground parts to their roots at a rate sufficient to supply the demand of root tissues. Adequate respiration of roots requires that the soil itself be aerated. This means that the soil pores must contain sufficient air, and must exchange gases freely with the external atmosphere so as to prevent a deficiency of oxygen from developing in the soil. Prolonged flooding or waterlogging of the soil impedes aeration, thereby inhibiting plant growth. Obviously, then, too much water is as bad as too little, and the entire soil-moisture balance must be maintained at a nearly optimal level continuously.

The amount of salt in the soil is another important determinant of plant growth. Water that is present in the soil is never chemically pure. Atmospheric moisture is of course distilled and essentially pure when first vaporized, but as it condenses to form clouds, and as it descends through the air as rain, it generally absorbs such atmospheric gases as carbon dioxide and oxygen, often with other gaseous products of our industrial activities. Along the seacoast, atmospheric moisture also picks up generous amounts of salt that enter the air as sea spray. Irrigation water, generally obtained from surface or subterranean reservoirs, frequently contains significant quantities of salts dissolved from rocks. Finally, while it is in the soil, the infiltrated water may dissolve additional salts contained in the soil itself. An excessive accumulation of salts in the soil causes a decline in productivity. "Soil salinity" is the term used to designate a concentration of salts in the soil that is harmful to crops. Along with waterlogging, it is a major hazard in semiarid and arid regions.

The strategy by which any particular type of plant adapts to its environment may depend not only on the conditions prevailing at a given moment, but also on an implicit set of expectations, so to speak, of what the future might bring. These expectations are not conscious but preprogrammed, as it were, into the plant's biochemical or biophysical responses. It is still a marvel just how different types of plant species come to have different strategies for survival, and how each can adjust its strategy in response to uncontrollable and often adverse conditions. On the whole, however, plants manage very well—better in general than animals, and certainly better than human beings, whom plants preceded and will doubtlessly outlast.

Suppose that plants of two species are growing side by side in competition. Now let us characterize one species as "optimistic" and the other as "pessimistic." The optimistic plants base their strategy on the assumption that the supply of soil moisture will remain plentiful, and hence only a small root system is needed to meet its water requirements. Therefore, the optimists invest their available resources (namely, their newly photosynthesized carbohydrates representing their growth potential) in forming more branches and leaves. With more green leaves, these plants might have a greater capacity for photosynthesis and therefore be able to produce more flowers, fruits, and seeds, thereby dominating the habitat. But what if these expectations prove to be wrong? Suppose the rains do not come, and soil moisture is limited. Then the greater transpiration resulting from the greater foliage of the optimistic plants soon exceeds the water-supplying power of the restricted roots, so these plants may end up being left, quite literally, high and dry.

a long drought. Hence they tend to invest their growth potential
in the preferential development of more roots rather than more
shoots. If their guess is wrong, these cautious plants will be
literally overshadowed by their more aggressive competitors.
If, on the other hand, their pessimistic expectations are borne
out, their curtailed foliar transpiration, coupled with their enhanced root system's ability to provide, will carry them through.
Then the meek shall indeed inherit the earth—at least for that
particular season.

Our exercise in hypothetical teleology is not to be taken
too literally, of course, as it was meant only to illustrate how
plant species can differ in adaptation to environments with varying
degrees of aridity.[3]

Plants that inhabit water-saturated domains are called hydro-
phytes, or aquatic plants. Plants adapted to drawing water from
shallow water tables are called phreatophytes. In contrast, plants
that can grow in arid regions by surviving long periods of thirst
and then recovering quickly when water is supplied, are called
xerophytes, or desert plants. Such plants may exhibit special
features, called xeromorphic, that are designed to store water
and minimize its loss; for example, succulent tissues, thickened
epidermis and a waxy surface cuticle, recessed stomates and reduced leaf area. Some plants are especially adapted to growing
in a saline environment and are called halophytes. It takes an
extreme degree of adaptability (or masochism) for certain plants
(xero-halophytes) to grow in an environment that is both dry
and saline at the same time.

Finally, there are plants that grow best in moist but aerated
soils, generally in semihumid to semiarid climates. Such intermediate plants are called mesophytes. Most crop plants belong in
this category. Mesophytes control their water economy by developing extensive root systems and optimizing the ratio of roots
to shoots, and by regulating the aperture of their stomates to
curtail transpiration during periods of water shortage. The latter
effect, however, necessarily entails restriction of photosynthesis.
Moreover, curtailment of transpiration reduces the evaporative
cooling of the plants, so they tend to warm up under the sun's

radiation, and as they warm up their rate of respiration rises. In effect, therefore, a thirsty plant consumes its own reserves and further reduces its growth potential. For these reasons, thirsty crop plants are generally much less productive than plants that are well-endowed with water throughout their growing season.

*These elaborately constructed forms, so
different and so dependent upon each other,
have all been produced by laws acting
around us.*

CHARLES DARWIN, *The Origin of Species*

7

THE TENUOUS
BALANCE

E COLOGY HAS TO DO WITH the interactive relationships among
living beings and their environment. More specifically,
the study of ecology is aimed at elucidating the laws by
which the various species and groups of species composing an
area's biological community are organized, and how they function
individually and jointly within their common environment. Such
a study can guide us in assessing how and to what extent humans
may utilize a particular environment without changing it irrevoca-
bly for the worse.

An interesting branch of general ecology, therefore, is human
ecology, the study of how humans have interrelated with the
earth. All civilizations have been, in one sense, adaptations to
the ecosystem. Yet in another sense, they have had ways of
utilizing and manipulating the ecosystem for the benefit of the
particular human society. That manipulation was occasionally
constructive, symbiotic with nature, and hence sustainable; more
often, it was destructive. Wherever the latter mode persisted
over an extended period of time, nature eventually exacted its
terrible revenge in causing the fall of those societies.

The classical analytical approach to the study of nature was to disassemble every system into components or phenomena that were to be studied in isolation, one by one; *ceteris paribus*—assuming all other things to remain equal, was the guiding presumption. That approach is inappropriate in attempting to understand the biosphere, where numerous interdependent phenomena occur either all at once, or in sequence, and all affect one another. Dynamic environmental systems therefore disobey the simplistic Euclidian postulate that the whole must equal the sum of its parts. In such systems, the whole is in effect greater than the sum of its separate parts, since it includes the gamut of their interactions.[1]

Only in recent decades did our community of specialists become fully aware that the field we attempt to define, understand, and manage is an open system, an integral part of a large, continuous and very complex environment. Hence what we do (or fail to do) in the way of water management, fertilizer or pesticide application, or waste disposal, may bring about repercussions within and beyond the field in ways we had scarcely expected and can hardly control. Our former traditional piecemeal approach to the system was not only inadequate but dangerous, as it led to oversimplified and erroneous conceptions about the workings of the system as a whole.

Enter the concept of an ecosystem, the ecologist's designation for nature's comprehensive organizational unit. It encompasses the host of plants, animals, and microorganisms that share a given domain and that might therefore influence one another through such modes as competition or symbiosis, predation and parasitism. It also includes the environmental complex of physical and chemical factors that influences those organisms and is often influenced by them. In every ecosystem one can distinguish among four groups of species: the primary producers—photosynthetic plants; the herbivores—plant-eating animals; the carnivores—animal-eating animals; and the decomposers—organisms subsisting on, and recycling, dead plant and animal tissues.

In a more or less stable ecosystem, all these groups interact in a manner that allows the entire composite population to thrive. A natural ecosystem is a self-regulating, self-optimizing, resource-

utilizing community in which the populations of the various
species are maintained in nearly steady proportions by an intricate
set of checks and balances. Any temporary disturbance, possibly
resulting from a chance deviation in climate (such as drought)
or an invasion by a new species, is damped out sooner or later,
so the ecosystem tends to restore itself and recover its former
stability. (To be sure, this is a simplistic "steady state" view
of ecology, ignoring long-term evolutionary developments as
well as the rapid transformations that may be triggered by sudden
environmental or genetic changes.)

It follows that the very nature of life on earth requires every
individual and species to adjust constantly, in order to reconcile
innate capacity and urge for growth with the opportunities and
constraints that arise in interactions with the environment and
with other beings sharing the same habitat.[2] A simple analogy
is that of a single species of bacteria introduced into a nutrient-
rich medium in a petri dish. An initial period of vigorous growth
occurs, but as the colony fills the limited space and uses up
the available nutrients, and as its waste products accumulate,
the initial bloom in bacteria is inevitably followed by stagnation
and then by decline and death. If other species are present and
are able to recycle the waste products, regenerating the nutrients
needed by the first species, then a symbiotic steady state may
develop, leading to a stable population of the two (or more)
species. The stability of such a steady state would depend on
the mutually complementary or competitive relationship among
the coexisting species, and on the maintenance of stable environmental conditions. In any case, none of the species would be
able to continue proliferating without disrupting the stability
of the community as a whole (the biome), thus ultimately destroying the basis for its own survival.

Our human species seems to have transcended those constraints as it has ranged over the entire earth and as it has
learned to manipulate the environment everywhere so as to create
conditions more advantageous to itself. So far, our success has
been phenomenal. But the story is not ended, and what has so
far seemed like success may yet set the stage for ultimate failure.

The chief agent of anthropogenic (man-influenced) transformation of the land has been agriculture. By its very nature, it is an intrusion and hence a disruption of the environment, as it replaces a natural ecosystem with an artificial one, established and maintained by man. The moment a farmer delineates a tract of land, separating it from the contiguous area by arbitrary boundaries and establishing it as his field, he is in effect declaring war on the pre-existing environmental order. Wishing to grow a particular crop (which may be of a species or a type not indigenous to the area, and therefore incapable of establishing itself there on its own), the farmer must now treat all the native species as noxious weeds or pests, to be eradicated by all possible means. However, in an open environment the wild species continue to reinvade their stolen domain, so the farmer's war is never finally won.[3]

The constant effort to prepare the field for seasonal planting and to eradicate weeds has traditionally involved repeated cultivations of the soil, often leading to excessive pulverization and compaction. Such mechanical manipulation tends to destroy the soil's natural aggregated structure and to render the soil surface particularly vulnerable to erosion.

Besides using mechanical means, farmers in recent decades have been relying increasingly on chemical pesticides to control weeds, pests, and diseases. This involves the application of a bewildering variety of both natural and—to an increasing extent—synthetic chemicals, and it charges the soil with potential and actual pollutants. Some of these substances are immobilized or degraded in the soil, but other toxic substances or their derivatives persist and eventually find their way into the biological chain. They may migrate by runoff and transported sediment to surface water reservoirs, by leaching through the soil to the groundwater, or by absorption into plants that are later consumed by animals.

All plants in general, and crop plants in particular, require certain essential mineral nutrients. The major elements are nitrogen, phosphorus, and potassium. Various additional "minor" elements such as sulfur, calcium, magnesium, and iron are required in smaller quantities. In natural ecosystems, the soil supplies nutrients to plants and thus sustains new growth by continuously recycling the residues of antecedent plant and animal life, including leaf litter, dead roots, and animal wastes.

In agricultural fields, some of the products of each season are harvested and removed from the field, rather than left to decay and return their nutrients to the soil. If left without proper replenishment, a cultivated soil tends to use up its initial reserves and to be depleted of readily available nutrients after a number of years, so a gradual loss of fertility typically follows the practice of cropping. Fertility can be maintained, however, by the regular addition of organic manures or of balanced mineral fertilizers. Modern agriculture relies on massive applications of chemical fertilizers to boost yields. As with pesticides, some of the residues of these fertilizers may migrate beyond the field and concentrate in groundwater or in surface water bodies, where they tend to cause eutrophication—a condition in which nutrient-rich waters induce the proliferation of algae and the consequent depletion of dissolved oxygen, much to the detriment of fish and other aquatic animals.

As long as agriculture was confined to small enclaves or limited tracts of land, while the greater continental area of the world remained undisturbed, the earth's environment as a whole was not seriously threatened. Degraded lands could be abandoned and new frontiers conquered, and still the earth's resources seemed limitless. But no more. Population growth has nearly filled all the empty spaces. Loss of natural fertility, erosion, waterlogging, salinization, pollution, and the annihilation of numerous species of plants and animals—such are the initially unforeseen but now global consequences of man's injudicious management of soil and water, in his unrestrained exploitation of the earth's resources.

PART III

THE

LESSONS

OF THE

PAST

We must, however, acknowledge,
as it seems to me, that man with
all his noble qualities . . . still
bears . . . the indelible stamp of
his lowly origin.

CHARLES DARWIN, *The Descent of Man*

8

HUMAN ORIGINS

ISTORY DOES NOT merely resurrect a dead past. In the words of Thucydides: "Knowledge of the past is an aid to interpretation of the future." If we can truly learn from past experience, we may be better able to improve our current use of the environment. If we focus our attention exclusively upon the predicaments of the moment, however, we may find ourselves repeatedly surprised by a host of bewildering problems seeming to come out of nowhere, without a past and hence without direction. How did these problems arise? Chances are, the seeds of the phenomena we witness today were planted some time ago by our predecessors, as indeed we are planting the seeds of the future—perhaps unknowingly—at this very moment.

The story of mankind begins more than three million years ago, when a genus of primates evolved to the point where it became recognizably humanoid.[1] Partly because of the baffling course of evolution itself, though, it is difficult to ascribe an exact age to humankind as it gradually diverged from its primate progenitors. Over extended periods of time, biological evolution appears to proceed very slowly by a long series of small, almost imperceptible, changes. Then, periodically, thresholds are reached that trigger seemingly sudden transformations. Such transformations may be due to chance occurrences of genetic mutations, or to shifts in environmental conditions, or—more likely—to combinations or sequences of these. Genetic and environmental changes may conjoin to trigger an unusually rapid preferential selection, and the consequent emergence, of a new biological type endowed with traits more advantageous than those of its predecessors. Analogous rapid transformations have also occurred in human cultural evolution.

Any attempt to describe the early course of humankind is thwarted by the fact that the very definition of what constitutes true humanity is somewhat arbitrary. Ever since Charles Darwin first elaborated on the possible circumstances of human origin in his 1871 book, *The Descent of Man,* anthropologists have been speculating on the sequence of events that gradually brought about the astonishing metamorphosis of a tree-dwelling, quadripedal, herbivorous ape into a ground-dwelling, bipedal, tool-making, omnivorous hominid. A crucial step appears to have been the shift from four-legged to two-legged locomotion (bipedalism). This was followed by further structural and functional evolution. The eyes were adapted to stereoscopic vision for judging distances. The hands, preconditioned to grasp branches with an opposing thumb, later developed a capability for the precision grip used in making and employing tools. All the while, the brain grew in size and function as it developed the ability to process more information and to generate complex logical thoughts.

Various hypotheses have been advanced regarding the origin of hominid bipedalism and all that followed. Such hypotheses, no matter how plausible, are virtually impossible to test or to prove conclusively. A long-held popular notion was that the evolutionary shift was induced by the need of an otherwise defenseless "ape" to make and use tools and weapons for hunting and for protection against predators. This notion accords with the idea that the principal effect of walking on two feet was to liberate the hands for the performance of tasks and the acquisition of skills that would have been impossible otherwise.

Other investigators view the origin of bipedalism as a means by which hominids could cover a larger territory in foraging for dispersed plant foods. This idea fits into the context of homi- nids occupying a more open environment than the dense forests to which primates were initially adapted. The environmental change may have been due to a shift of climate—leading to a partial drying of the original habitat—that apparently took place during the later stages of the Cenozoic era. According to this view, developing the facility to walk and run over the land on two legs, with an upright posture allowing a longer view of the landscape, was an ape's adaptation to living where apes do not normally live. Still others suppose that bipedalism developed for long-distance trekking to scavenge from migrating ungulate herds, like those now found in the Serengeti Plains of Tanzania. This supposition is consistent with the recent perception that, in addition to gathering plant products, the very early hominids might have engaged more in scavenging than in hunting as a means of subsistence. Whatever motivated or triggered the transition to bipedalism, it proved to be irreversible, and its ultimate consequences were fateful for the subsequent course of humankind.

Our species' birthplace was apparently in the continent of Africa, and its original habitat was probably the subtropical savannas which constitute the transitional areas of sparsely wooded grasslands lying between the zone of the humid and dense tropical forests and the zone of the semiarid steppes. We can infer the warm climate of our place of origin from the fact that we are naturally so scantily clad, or furless; and we can infer the open landscape from the way we are conditioned to walk, run, and gaze over long distances.

Fossil discoveries in East Africa during recent decades have revealed facts that have added dramatically to our knowledge of human origins. Skeletal finds suggest a succession of primate and hominid types starting several million years ago and progressively approaching the structure that is definitely characteristic of humans. Evidence seems to suggest that the beginnings of stone tool-making followed the origin of bipedalism by more than a million years. The earliest known hominid capable of a striding bipedal gait and a precision grip (circa 3.75 million years ago) was discovered in Tanzania and in Ethiopia, and has been named *Australopithecus affarensis*. Fossils of one of its presumed descendants, a tool-making hominid called *Australopithecus africanus,* were found in deposits dated some 2.5 million years ago in Sterkfontein cave in South Africa. In time, the tools made by hominids had developed into distinct, consistent implements for cutting, scraping, and grinding foods, including plant products such as nuts and grains, and animal products (flesh, skins, and bones). Such implements were needed to compensate for the inherent inadequacy of hominid teeth and jaws to support the changing life style of the wide-ranging animal that eventually evolved into the genus called *Homo.*

For at least 90 percent of its career, the human animal existed merely as one member of a community of numerous species who shared the same environment. Humans were adapted to subsist within the bounds defined by the natural ecosystem: they neither dominated other species nor brought about any fundamental modification of the common environment. By and large, our ancestors led a nomadic life, roaming in small bands, foraging wherever they could find food. They were gatherers, scavengers, and hunters. Unlike their primate cousins who remained primarily vegetarian, humans diversified their diet to include the flesh of whichever edible animals they could find or catch, as well as a variety of plant products such as nuts, berries and other fruits, seeds, and some succulent leaves, bulbs, tubers, and fleshy roots.

The story of how humans ascended from their humble apelike origins to venture far from their birthplace, and range over a variety of climates and landscapes, is a remarkable saga of audacity, ingenuity, perseverance, and adaptability. In fact, humans have proved to be the most adaptable of all terrestrial mammals. Their mode of adaptation was not entirely genetic or physical: there was not enough time for that. Rather, their adaptation was in large part behavioral. Instead of relying on physical prowess, they had to use inventiveness to survive the elements and to compete successfully against stronger animals. In the course of their migration and expansion, our ancient forebears therefore had to develop and mobilize all the cunning and intelligence that eventually made them—and us—so unique a species. The increase of brain size and manual dexterity, as well as the invention of various stratagems, gradually enabled humans to overcome the constraints of their ancestry.

By 1 million years ago, hominids had become taller (about 1.5 meters in height), and had acquired a larger brain. Some time later, so-called *Homo erectus* had learned to make and use fire, probably at first only for cooking and softening food. That achievement, following upon the development of stone tools, was a momentous technical innovation, celebrated in the Greek myth of Prometheus. Eventually, it had a great effect on the environment. Some evidence has been found in Southern and Eastern Africa of repetitive occurrences of brush fires, apparently set by humans nearly a million years ago. This early manifestation of pyrotechnology, whether purposeful or accidental, signifies the beginning of human manipulation of the earth's ecosystems. The use of fire became even more important when humans moved out of the tropics into colder climes, where bonfires and hearths were needed to warm their shelters in winter.

By about 250,000 B.P. (Before the Present), humans had evolved into the type that anthropologists call *Homo sapiens,* and had spread to Europe and Asia. Though this geographic migration could not have been a consistent expansion, as it must have been influenced by the alternating glacial and interglacial cycles of the Pleistocene age, it eventually spread humans throughout those continents. (There is no evidence that people had arrived in the Americas, or in Australia, until about 40,000 B.P.) Some time before 50,000 B.P., a race of humans called Neanderthals, who lived during the last Ice Age, were making cutting tools with flaked flint. By about 40,000 years ago, modern humans (*Homo sapiens sapiens*), evidently indistinguishable from us today in physical features and in intelligence, had gained dominance.

Clad in sewn garments made of animal skins, able to make

and use a variety of implements, and armed with a growing array of weapons—including spears and bows and arrows—humans were able to range and settle in locations and climes far from their ancestral home. All the while they continued to evolve biologically through genetic change and natural selection, increasingly aided by cultural and technological development. To survive the harsh winters of colder climates, they had to find or construct shelters, and to huddle in family or tribal groupings for mutual assistance and the rearing of their slow-growing offspring. In their leisure time, they painted animals on cave walls and carved ritual objects. They also had to contrive increasingly sophisticated methods of obtaining and storing foods, including the selective gathering, processing, and preservation of biological products, and eventually the domestication of plants and animals.

This series of changes has been termed the Paleolithic (Early Stone Age) Transformation.[2] It was marked by the development of adaptive mechanisms for recognizing and exploiting potentialities within the environment. Utilizing and further refining their distinct physical, intellectual, and social abilities, our ancestors increasingly set themselves apart from other species of animals. Gradually, as they continued to elaborate and perfect their tools of wood, bone, and stone, as well as their techniques and social organization, humans assumed an increasingly active and eventually dominant role in shaping their environment. Each modification of the environment entailed additional human responses, which in turn further modified the environment, so that a process of escalating dual metamorphosis was instigated. Human intelligence and culture were both cause and effect in that fateful interplay. The peculiarly dynamic and progressive evolution of human ecology is the true history of our species.[3]

At some point, humans began to use fires deliberately and systematically to flush out game and to modify the vegetation. The resultant suppression of woody plants and the fertilizing effect of ash encouraged the growth of herbaceous plants and improved their nutritional quality. This benefitted foraging species and raised the carrying capacity for game animals. It also facilitated foot travel and hunting by humans. In time, the also set the stage for the advent of agriculture.

The practice of burning vegetation, along with the increasing skill of humans as hunters, may have contributed to the extinction of several large herbivores, which had no effective defense against their fire-setting and weapon-wielding two-legged predators. In North America, for example, two-thirds of the mammalian megafauna (species with adults weighing 50 kilograms or more) present at the end of the Pleistocene era (circa 11,000 B.P.) disappeared, including 3 genera of elephants and 15 of ungulates. In Eurasia, the losses included the woolly mammoth, woolly rhinoceros, giant Irish elk, musk ox, dwarf elephant, and steppe bison. It is impossible to state definitively, however, to what extent these extinctions may have been caused (or affected) by climatic changes. In Northwestern Europe, the same practice of forest burning is suspected of having resulted in the development of heathlands and bogs. Areas in North America also seem to have been fire-managed by pre-Columbian Indians.

In Australia, the intentional maintenance of grasslands and open woodlands by periodic burnings was a regular practice of the hunter-gatherer aborigines. In the Cape York Peninsula, there is evidence that the aborigines used repeated firings to eradicate the original vegetation and to encourage the preferential growth of cycad trees, which yield edible kernels. Further into the interior of Australia, the aborigines used fire for hunting, land clearing, communication, and domestic purposes. The animals flushed out by the flames could be captured more easily. In the southern part of Australia, a high fire-frequency apparently helped to convert the original climax forest* of beech trees into a heath or tussock grassland.[4]

As vegetation is affected by fire-setting hunters, so are soils. Following repeated fires and deforestation, soil erosion and landslides often result in the greatly increased transport of silt by

---

*A climax forest is a community of trees and associated species that has attained stability (equilibrium) within its environment. This is, of course, not an absolute definition, as in time climates shift, species evolve, and the environment changes.

streams, and in the deposit of that silt in river valleys and estuaries. The dating of fluvial sediments in river valleys in England, for example, suggests that they were the products of erosion caused by anthropogenic clearings in the originally closed deciduous forest during the Late Paleolithic period.

The fact that pre-agricultural people caused substantial changes in their environments does not necessarily imply that they were *always* destructive. Not all changes are inevitably deleterious—only those that create unsustainable conditions and result in progressive degradation. The mere substitution of one type of vegetation for another may even be beneficial in the long run, provided the new landscape is more productive and at least equally sustainable. The problem, however, is that it is ever easier to set fire to dry vegetation than to predict, let alone control, the consequences of the resultant conflagration, which is likely to be destructive if repeated too often.

The gradual intensification of land use continued throughout the Paleolithic period, so that by its later stages nearly all the regions of human habitation had experienced some anthropogenic modification of the floral and faunal communities. At some stage, humans began to delineate sections of the environment which they could control and manage to suit their special needs, and in which they could find convenient and secure shelters for at least temporary habitation. They recognized nutritional and medicinal plants, observed their life cycles, and learned to encourage and take advantage of their natural propagation patterns. They learned to build rafts and boats of various types and thereby to exploit aquatic resources. As they became more mobile, the rivers and lakes that were once barriers became arteries of travel and transport. They developed implements for grinding and cooking vegetable and animal products, and weapons for hunting larger game animals. Success in these endeavors provided them with the leisure to develop social and cultural activities: music, dances, rituals, ceremonies, storytelling, rites of passage, creative arts, and the crafting of useful and decorative articles. Their success also brought about a growth in population, which in turn induced further geographic expansion and intensification of land use in quest of additional sources of livelihood.

In toil shalt thou eat of the earth all the days of thy life.

GENESIS 3:17

9

THE AGRICULTURAL TRANSFORMATION

THE BIBLICAL STORY describing the banishment of Adam and Eve from the Garden of Eden may be taken to symbolize humanity's transformation from the carefree "child of nature" hunting-gathering-wandering phase of existence to a life of toil and responsibility as permanently bound tillers of the soil. The actual initiation of settlement appears to have begun in the Late Paleolithic (sometimes called Mesolithic) period that preceded the advent of farming by several thousand years. On finding a particularly favorable location, a clan of humans would naturally tend to prolong its stay there so as to take advantage of its favorable conditions. Those conditions might include an assured supply of water, a relative abundance of game or of edible plant resources, access to useful raw materials such as flint or wood, a benign climate or shelter against inclement weather, as well as safety or protection against potential enemies.

The process of intensification of land use can be seen as an adaptation to increasing population pressure. Several millennia of occupation by hunter-gatherers, even at a very low density and slow rate of population growth, could have filled up the terrain and decimated the natural forageable resources to the point where subsistence could become difficult.[1] The choice would then be between migration and some form of intensification aimed at inducing the same area to yield a greater supply. Free hunting would be supplanted by manipulative hunting, based on the use of fire to modify the vegetation, or of various stratagems to lure and trap a greater number of animals. The next step would be the selective eradication of undesirable species and the encouragement of desirable ones, leading eventually to herding and domestication. Similarly, selective manipulation of plant communities would involve suppressing some species and promoting the growth of others. The entire series of activities would quite logically lead to plant domestication and propagation, and to purposeful land and soil management aimed at creating favorable conditions for crop production—that is to say, these activities would culminate in the development of agriculture and the agricultural way of life.[2]

The Agricultural Transformation is very likely the most momentous turn in the progress of humankind, and many believe it to be the real beginning of civilization.[3] Often called the Neolithic Revolution, this transformation apparently first took place in the Near East between 10,000 and 8,000 years ago, and was based on the successful domestication of suitable species of plants and animals. The ability to raise crops and livestock, while resulting in a larger and more secure supply of food, definitely required attachment to controllable sections of land, and hence brought about the growth of permanent settlements and of larger coordinated communities. The economic and physical security so gained accelerated the process of population growth, and necessitated further expansion and intensification of production. A self-reinforcing and self-perpetuating pattern thus developed, so the transition from the nomadic hunter-gatherer mode to the settled farming mode of life became in effect irreversible.[4]

Compared to the long period of two or more million years during which our ancestors were hunters and gatherers, the brief interval of two thousand or so years required to accomplish the Agricultural Transformation over most of the region known as the Near East seems almost instantaneous. But why did humans suddenly give up their long natural existence as hunters and gatherers, to which they were so thoroughly adapted by evolution, both physically and culturally? What impelled them to join together in larger and larger groups, thus presaging the densely packed and often unhealthful cities that ultimately became the characteristic mode of life in much of the modern world? How did the sedentary life become so universally appealing that it was so quickly adopted by people in practically all regions of the world? And why did this momentous transformation first take place in the Near East of all regions? What was the natural setting in which the fateful change was initiated? These are questions to which we still have only partial answers.

Clearly, the old stereotypic portrayal of the Late-Paleolithic pre-agricultural people as ignorant savages is erroneous. We have much evidence, both historical and derived from present-day hunter-gatherers, to prove that their understanding of the environment within which they lived was sophisticated indeed. No doubt they knew a great deal about the life cycles of plants and animals, for instance, as their livelihood depended on that knowledge. In a real sense, therefore, they were professional botanists and zoologists.

Contemporary or recent hunter-gatherers, such as the Bushmen of Southern Africa, the Panare of Amazonian Venezuela, the Dinka of the Sudanese Sudd, and the aborigines of Australia, still maintain and utilize the rich lore amassed by countless generations of their forebears. They know not only how to distinguish nutritious plants from those that are non-nutritious or poisonous, and how to detoxify harmful vegetable products, but also how to use plant-derived drugs, narcotics, arrow poisons, gums and resins, glues, dyes and paints, as well as fibers for spinning ropes and for weaving mats, baskets, and cloth. Thus the reason they did not, for so long, choose to take up agriculture is either that they had no need to do so or that local conditions were not conducive. As long as the population remained low enough so that the carrying capacity of the habitat was not exceeded, humans could continue to subsist as gatherers and hunters and were under no compulsion to change their traditional mode of life.

Some anthropologists and prehistorians have argued that

semi-nomadic hunting and gathering in small bands was an easier and healthier lifestyle than permanent farming, so the transition to the latter may actually have been disadvantageous, rather than immediately advantageous as it has often been portrayed. Reliance on farming imposed a monotonous diet of grain and a few other edible crops, instead of the rich and varied nutrition which could be obtained by hunting and gathering. Furthermore, life in larger groups residing in dense settlements increased the incidence and spread of contagious diseases, and may thus have shortened the average longevity. The contrary and still the more prevalent view is that the advent of agriculture ensured a supply of food, and freed humans from the need to roam endlessly over the countryside in search of edible wild plants and animal prey. Moreover, since the early farmers domesticated animals as well as plants, their diet may have been no worse, and in some cases better, than that of hunter-gatherers. Finally, stable communities provided more secure conditions for rearing children.

Notwithstanding the arguable disadvantages of the original Agricultural Transformation, the fact remains that this change did occur, that it was rather rapid, and that it was essentially irreversible. Hence, *ipso facto,* it must have been advantageous overall, though it certainly created its own problems. There must have been something in the condition of humans that impelled that transformation once it became possible. That something may well have been an antecedent increase in human population density following the use of tools, weapons, and techniques that had increased the efficiency of hunting and gathering to the point where human groups were depleting the supply of game animals and edible wild plants within the areas available to them.

The advent of farming itself could not have been a sudden discovery or invention by some individual genius. Rather, it must have been the culmination of a long series of observations and trials by numerous generations of humans transmitting and augmenting their experience and methods, until the knowledge, technology, and circumstances were ripe for the seminal transformation.

Although agriculture seems to have been developed first in the Near East, that region is by no means the sole center of crop and animal domestication. At different stages, separate and very likely independent developments took place in other centers, each with its own selection of crops. Among these centers are Sub-Saharan Africa, East Asia (China), Southeast Asia and Oceania, and the Americas.

The process of plant domestication and the evolution of crop plants from their wild progenitors is a fascinating topic of study, made progressively more difficult by the globally accelerating destruction of natural habitats and of native plant communities. By domesticating plants and developing crops, humans created biological artifacts that could no longer thrive autonomously without constant care.[5] Reciprocally, humans had become so dependent on their crops that, in effect, their crops had domesticated them.

The domestication of animals occurred as a consequence of hunting, not necessarily in conjunction with the domestication of plants. Consequently, the herding of animals and the husbandry of crops were in some places complementary, and in other places divergent, activities. The benefits of animal domestication were obvious—secure supplies of meat, milk, fur, leather, wool, and even bones and horns for tool-making. Animal manure could serve to fertilize crops. Larger animals could also assist in the performance of laborious tasks and in transportation. However, the cost in terms of human labor was high. Human herders needed not only to feed and breed their animals and to confine them to prevent their escape, but also to protect them against predators, diseases, and climatic vagaries. This required a level of planning, commitment, and consistency never before undertaken by humans. Consequently, both the domesticated animals and their keepers developed a mutual dependency.

Pastoralists were able to exploit niches marginal to the agricultural zone, like patches of scrub and grass at the edges of fields and paths, as well as semiarid hill lands peripheral to the river valleys that became the centers of cultivation. Such extensive utilization of patchy and seasonal pastures required moving the animals periodically from one place to another, either from a

permanent base, or by moving the human abode along with
the animals—a mode called transhumance. The roving pattern
of grazing could become especially extensive in drought-prone
regions, where the sparse growth of forage, and the paucity of
water, require graziers to roam almost constantly in search of
sustenance for their flocks, thus assuming a nomadic life.

As agriculturists, human beings began to affect their environment to a greater degree than ever before.[6] They cleared away
the natural flora and fauna from selected tracts, and in their
place introduced and nurtured the species or varieties of plants
and animals they preferred. By so doing, they modified the
natural ecosystems of increasingly large areas, until they eventually altered entire regions. Their success, as measured in terms
of population growth, was considerable, but this success sometimes resulted in the practically irreparable degradation of the
once-bountiful environment in which agricultural development
began.

The Agricultural Transformation radically changed almost
every aspect of human life. Food production and storage stimu-
lated specialization of activities, and greatly enhanced the division
of labor which had already started in hunting-gathering societies.
The larger permanent communities based on agriculture required
new forms of organization, both social and economic. Domestication undoubtedly affected family structure and the roles and
status of men, women, and children. With permanent facilities
such as dwellings, storage bins, heavy tools, and agricultural
fields came the concept of property. Specifically, private ownership
of land may well have originated with the advent of agriculture.
So also might have the private ownership of springs and other
water resources. The inevitably uneven allocation of such property
resulted in self-perpetuating class differences. Religious myths
and rituals, as well as moral and behavioral standards, developed
in accordance with the new economic and social constellation
and the new relationship between human society and the environment.

The evolution of agriculture has left a strong imprint on
the land in many regions. The vegetation, animal populations,

slopes, valleys, and soil cover of land units have all been altered.
The processes of tillage and fallowing, of terracing, of irrigation,
and drainage have had considerable consequences for such processes as the erosion of slopes and the aggradation of valleys,
as well as the formation of deltas in seas and lakes where silt
from the land surface naturally comes to rest. Soil lost from
deforested and subsequently cultivated slopes is unlikely to be
regenerated unless the land is allowed to revert to its forest
cover for many scores, perhaps even many hundreds, of years.

Pastoralism, as well as cultivated farming, can cause a great
deal of environmental damage. During dry seasons, when large
numbers of animals are kept on pastures least able to sustain
them, the land is denuded of its vegetation and made most
vulnerable to the erosive onslaught of winds and of violent rainstorms that may occur at the end of the dry period. If overgrazing continues over a long period of time, the environmental
damage can be profound. In antiquity, shepherds in the fringelands of the Mediterranean region were notorious as plunderers
of land. Though they must have tried, as do present-day pastoralists, to maintain a rough equilibrium between stocking rates
(the number of animals grazed on a unit area of pasture) and
the average carrying capacity of the range, that equilibrium
could only be maintained as long as the range remained more
or less stable. However, such a system would naturally break
down during periods of drought, when the pressure on the shriveled vegetation would soon become excessive. To survive during
such periods, the pastoralists would have had no recourse but
to invade the land of the neighboring farmers. The ancient enmity
between these groups has long been legendary and implacable,
as it still is today in some semiarid regions.

Two or three millennia after the initial Agricultural Transformation, there began a further process of fundamental change;
namely, the process of urbanization.[7] It was made possible by
the very success of agriculture, as the people involved directly
in farming produced surpluses beyond their subsistence needs.
These surpluses could then support the artisans, traders, priests,
administrators, and kings who resided in the cities. The develop-

ment of cities was not merely an increase in the size of settlements, but a qualitative change in the structure of society and its relationship to the environment. Today most of us belong to urbanized societies and live in cities quite detached from the land and its natural ecosystems.

The artificial environment of our cities owes many of its features to the early cities developed five thousand years ago in lowland Mesopotamia, and then elsewhere in the Near East. Among numerous innovations attributable to these early cities are writing, formal codes of law, political and ecclesiastical hierarchies, craft specialization, monumental art, mass-production industries, metallurgy, mathematics, scientific and engineering principles, architecture, large-scale trade, and organized warfare in the form both of massive defensive fortifications and long-distance offensive campaigns. The scale and intensity of land and water management in the agricultural hinterlands serving the cities had to change accordingly.

A land of wheat and barley, vines,
fig trees and pomegranates,
A land of oil olives and honey;
A land wherein thou shalt eat
bread without scarceness.

DEUTERONOMY 8:8–9

10

EARLY FARMING IN
THE NEAR EAST

THE PROCESS OF DEVELOPING a dependable food-producing system was a complex sequence of steps, starting with an initially extensive gathering economy that tended to become increasingly intensive, and culminating in a complete revolution in human society and its management of the environment. An essential step in that process was the selection of favorable wild plants in their natural habitats and their domestication and transformation into artificially propagated crops, to be grown at will in areas that might be far removed from their place of origin.

The end of the Pleistocene and the beginning of the Holocene era (some 10–12 thousand years ago) was a time of great climatic transition. The last ice age ended and a warming trend prevailed. Areas that had been cold and inhospitable in centuries past burst forth with a profusion of plants and animals that responded to the longer and warmer growing seasons. Having survived the vicissitudes of the ice age, doubtlessly thanks to their growing ingenuity and acquired skills, humans now found themselves in a more auspicious ecological situation, in which they could not only survive but even prosper and multiply.

In the Near East, they found a particularly favorable region

71

for subsistence and habitation. Evidence of this early habitation, the so-called Natufian culture (12,000 to 10,000 B.P.), has been found by archaeologists in the hills of modern Israel. The Natufians were apparently the first hunter-gatherers to make the transition to permanent settlement.[1] Though they continued to live off the native (albeit modified) environment rather than cultivate crops, they were apparently the forerunners of the earliest farmers. The Natufians built elaborate stone houses, had food preparation areas with mortars and pestles, and maintained storage facilities for the wild grain that they collected. Numerous potential crops grew wild on these relatively humid mountains, hills, and valleys. As people gathered the edible grains, fruits, nuts, stalks, leaves, or bulbous roots of these various plants, they observed their mode of growth and learned much about their propagation.

Prominent among the native plant resources of the Near East were wild species and varieties of the graminea (grass-related) and leguminosa families, whose seeds could be collected and stored to provide food for several months.[2] Most native plants scatter their seeds as soon as they mature, and are therefore difficult to harvest efficiently. A few anomalous plants, however, due to chance mutations, retain their seeds. The discovery and preferential selection of such seeds, and their propagation in favorable plots of land, constituted the real beginnings of agriculture, providing the early farmers with crops that could be harvested more uniformly and dependably than could the wild plants.[3]

The most important of the early crop domesticates were the annual cereal grains: barley and especially wheat, along with various leguminous grains, such as lentils, peas, chickpeas, and vetch. As settlements and villages acquired permanence, several fruit-bearing trees (which require years to mature) could also be domesticated. These included figs, olives, and dates, as well as grapes, pomegranates, and almonds. The earliest animal domesticates were sheep, goats, pigs, dogs, and cattle.

The progenitors of the region's cereal crops—namely, wild emmer wheat, wild einkorn wheat, and wild barley—evidently originated in the broad arc of uplands and foothills fringing

not any historic shift in climate might have occurred since the
beginning of the Holocene and might have affected the geographic
distribution of these species, it is interesting to note that stands
of these plants are prevalent even today in the hills of northern
Israel, Lebanon, western Syria, southern Turkey, northeastern
Iraq, and western Iran. Patches of these wild cereals would surely
have constituted an attractive source of food for pre-neolithic
hunter-gatherers. The wild grain could easily be harvested with
the flint-bladed sickles of the period. Native patches or stands
of these cereals were naturally limited in size, so the people
dependent on the grain would obviously wish to extend such
stands, by actively helping to spread their seeds and by selectively
eradicating competing vegetation.

As long as human intervention was confined merely to harvesting the wild grain, the effect would have been to encourage
such wild-type characteristics as shattering rachis (the spikes
connecting the seeds to the stalk) and nonuniform maturation,
since it was the seeds that escaped the harvester that tended to
produce the next spontaneous generation of wild grain. In contrast, the harvested batch of seeds would be selected in favor
of non-shattering and uniform maturation. As soon as humans
began to sow the seeds that they had harvested, they automatically—even if unintentionally—initiated a process of selection
in favor of the non-shattering genotype. Each season, most of
the seeds that shattered evaded the harvest, while most of the
seeds that remained attached were harvested and hence tended
to concentrate in the seedstock disproportionately to their preva-
lence in the wild. Similarly, the seeds that matured early were
shed before the harvest, and those that matured late were unripe
at the time of harvest, so neither contributed to the seedstock.
By this process, the proportion of the non-shattering, uniformly
maturing genotype was enhanced progressively until it became
a dominant characteristic of the crop. Consequently, of all the
adaptations that distinguish domesticated crops from their wild
progenitors, the non-shattering and uniform-ripening traits are
the most conspicuous.

The initial efforts at domestication probably took place close to each crop's center of origin,[4] which in the case of the cereal grains would have been the uplands and foothills girding the Fertile Crescent. The early farmers would naturally tend to seek a favorable plot of ground from which they could remove competing plants and in which they could conveniently sow their seeds with reasonable expectation of a worthwhile yield. Such plots were likely to be located in intermontane valleys, where the ground is relatively level and the alluvial soils are generally deep and fertile. Remnants of small Neolithic field plots have been found in some of the narrow valleys of the Carmel and Galilee ranges in Israel.

An important factor in the evolution of agriculture in the Near East, as elsewhere, was the development of the tools of soil husbandry. Seeds scattered on the ground are often eaten by birds and rodents, or subject to desiccation, so their germination rate is likely to be very low. Given a limited seed stock, farmers would naturally do whatever they could to promote germination and seedling establishment. The best way to accomplish this is to insert the seeds to some shallow depth, under a protective layer of loosened soil, and to eradicate the weeds that might compete with the crop seedlings for water, nutrients, and light.

The simplest tool developed for this purpose was a paddle-shaped digging stick, by which a farmer could make holes for seeds. The use of this simple device was extremely slow and laborious, however, so at some point the digging stick was modified to form the more convenient spade, which could not only open the ground for seed insertion but also loosen and pulverize the soil and eradicate weeds more efficiently. In time, the spade developed a triangular blade, initially made of wood but later made of stone, and eventually of metal. Such a spade, initially designed to be used by one person, was later modified so that it could be pulled by a rope so as to open a continuous slit, or furrow, into which the seeds could be sown. A second furrow could then be made alongside the first, to facilitate seed coverage. In some cases, the rows were widely enough separated

to permit a person to walk between the rows, weeding the
cultivated plot.

The man-pulled traction spade or *ard* gradually metamorphosed into an animal-drawn plow. The first picture of such a plow, dating to 3000 B.C.E., was found in Mesopotamia, and numerous later pictures have been found both there and in Egypt, as well as in China. It was not long before these early plows were fitted with a seed funnel, so that the acts of plowing and sowing could be carried out simultaneously. The same ancient implement is still very much in use today throughout the Near and Middle East.[5]

Although the development of the plow represented a huge advance in terms of convenience and efficiency of operation, it had an important side effect. As with many other innovations, the benefits were immediate, but the full range of consequences took several generations to play out, long after the new practice became entrenched. The major environmental consequence was that plowing made the soil surface—now loosened, pulverized, and bared of weeds—much more vulnerable to accelerated erosion. In the history of civilization, contrary to the idealistic vision of the prophet Isaiah, the plowshare has been far more destructive than the sword.

Though perhaps slower than the effects of land clearing for cultivation, the results of herding and overgrazing are ultimately no less destructive. In addition to being the natural habitats for the wild progenitors of several of the principal cultivated grain grasses, the mountains, foothills, and valleys of the Near East also hosted wild sheep, goats, pigs, and cattle that were later domesticated. These animals are at home in ecotonal habitats where grassland, brushland, and forest interpenetrate. Here, intensified herding, especially during drought seasons, eventually became a force for the destruction of the natural vegetation on which it had originally depended. Goats not only browse their favorite shrubs but can climb right up into trees to eat the foliage, and they eagerly consume trees seedlings, so that where they are constantly herded, forests cannot regenerate. Sheep, too, can do great damage when they overgraze, since they will eat grass, roots and all, and their sharp hooves, like those of goats, tear up the sod and pulverize the soil. Cattle, though not quite so destructive, can also overgraze, and herders often set fires to encourage the growth of grass.

As the early farming venture met with some success, the activity spread and the growing farming population could no longer be confined to the narrow intermontane valleys where agriculture apparently began. Villages were formed in the larger valleys below the foothills, and along the coastal plains of the Near East. For quite some time, villagers evidently combined localized farming with continued gathering activities and hunting forays. Increasingly, however, they became attached to their farming sites. More and more, their artifacts (grinding stones, stone mortars and pestles, ground-working and planting implements) and their installations (grain-storage pits or bins, and animal corrals) became permanent and non-transportable. Thus, the process of sedentarization, which actually began some time before the advent of agriculture, was reinforced by the vocation of farming. With permanent habitation, an important new industry could be developed—pottery, which began in the Near East about 8,000 B.P. The shaping and baking of clay to form hardened vessels for grain, for liquid storage, and for cooking, represented the first transmutation of matter by humans. Such an innovation could not have been possible, owing to the fragility of the ceramic objects, during the nomadic phase.

The Mediterranean-type climate of the Near East is at best semihumid, but more typically semiarid, with a rather high incidence of drought. Hence the practice of rainfed farming could not provide anything like total food security. The early farmers who depended only on seasonal rainfall to water their crops were always at the mercy of a capricious and highly unpredictable weather regime. The Hebrew Bible, for instance, is replete with references to the ever-present threat of drought and consequent famine. In time of need, therefore, it was only logical for farmers located near river courses to attempt to augment the water supply to their crops by artificially conveying water from the river—first by hand, and later by digging a diversion

channel. It was also logical to try to raise crops on riverine
flood plains that were naturally inundated, and thereby irrigated,
periodically.

At some point, then, farming was extended from the relatively
humid centers of its origin toward the extensive river valleys
of the Jordan, the Tigris-Euphrates, the Nile, and the Indus.
As the climate of these river valleys is generally quite arid, a
new type of agriculture based primarily or even entirely on irrigation came into being. With a practically assured perennial water
supply, an abundance of sunshine, a year-round growing season,
deep and fertile soils, and relative security from the hazards of
drought and erosion that beset rainfed agriculture, irrigated agriculture became a highly productive enterprise. However, behind
its success lurked an insidious problem which could not initially
have been foreseen: the problem of land degradation.

A river rose in Eden to water the garden.

GENESIS 2:11

11

SILT AND SALT IN MESOPOTAMIA

THE SUCCESSES AND FAILURES of past societies emphasize the crucial importance of the soil-water system in determining the long-term viability of all civilizations. Poor management, whether rooted in ignorance or indifference, can be ruinous. Nowhere is this principle more tragically apparent than in southern Mesopotamia. Once a thriving land of lush fields, it is now largely desolate, its great cities now barren mounds of clay rising out of the desert in mute testimony to the bygone glory of a spent civilization.[1]

In the second half of the fifth millennium, B.C.E., when the art and the benefits of farming were already widely established in the Near East, a group of people of uncertain ethnic and geographic origin began to colonize the lower courses of the Tigris and Euphrates rivers. Their neighbors to the north called them Sumerians; they called themselves "the dark-headed," and referred to their country simply as "the land." Sumer was a flat plain of brown alluvium, dusty when dry and miry when wet, swept by desiccating desert winds (the Babylonians, as well as the Egyptians, believed that the sky had been separated from the earth by the wind), and deluged periodically by the

78

sudden devastating overflows of the twin rivers. In Mesopotamia,
the myth of the great flood unleashed by God to scourge the
sinful—related in the Gilgamesh epic, which may have preceded
the Biblical account—was anchored in bitter experience.[2]

The first settlements were clumps of huts in marshes alongside
the watercourses of the lower Euphrates, much like the reed
huts of present-day marsh dwellers who still inhabit these areas.
Through diligence and ingenuity, the Sumerians gradually transformed their perilous land from a seasonally parched plain with
interspersed swamps, periodically or spottily too dry and too
wet, into a land of extensive grain and forage fields and date-
palm plantations. A Sumerian myth refers to the introduction
of cereals from the distant highlands. Another Sumerian myth
describes farmers as "men of dikes and canals." Those farmers
could feed ten times as many people as farmers on equal plots
without irrigation. The canals served as waterways and fishponds;
they watered palm groves and grasslands which furnished feed
for sheep and cattle, as well as grain fields of legendary fertility.
(Herodotus heard that in Mesopotamia grain yielded two- and
three-hundredfold the amount of seed planted. This is undoubtedly a great exaggeration; even a thirtyfold yield is impressive, and much more plausible.) The Sumerians traded the transportable products of their husbandry for the raw materials—
such as building stones, metals, and gems—that were lacking
in their land.

The surplus production of their farmers enabled the Sumerians
to develop the world's earliest urban society. The first Sumerian
cities were small mud settlements. Later, houses were built of
mudbrick joined by bitumen. Later still, the Sumerians used
burned bricks to build their homes and temples. The latter
were towers built on multistoried terraces, referred to in the
Biblical story of the tower of Babel challenging the heavens.[3]
Among the Sumerian cities were Eridu, the southernmost and
probably the earliest of all, Uruk, Nippur, Ur, Lagash, and
Kish.

In this age of the first cathedrals, probably around 3,300
B.C.E., the Sumerians made their greatest contribution to the

advance of civilization: the invention of writing. Using clay as their writing material, the Sumerians wrote with a reed stylus, the end of which could be pressed into the soft clay so as to make a series of wedge-shaped marks (a script called cuneiform, after the Latin word *cuneus,* meaning wedge). The patterns formed by those marks were schematic representations of objects used to signify syllables, objects, or ideas. When sun-dried or fired, such clay tablets became practically indestructible records. Thousands of them have been discovered by archeologists, and many of those that have been deciphered pertain to agriculture.

We owe much to the Sumerians. They developed sailboats, wheeled vehicles, the potter's wheel (the first industrial machine with continuous rotary motion), yokes for harnessing animals and animal-drawn plows, weighted levers for lifting water (called *shadoofs* in the modern Near East), accounting procedures, literature (including epics and love songs), and lawbooks. They also developed fortifications, weapons, war machines, and an entrenched bureaucracy.

Notwithstanding their achievements, the Sumerians did not last. As they began to decline, during the third millennium B.C.E., the Akkadians, who spoke a Semitic language, gradually learned the ways of the Sumerians and eventually superseded them. Toward the middle of the twenty-fourth century B.C.E., Sargon I ("the Great"), an Akkadian minister of the Sumerian king of Kish, gained control over Sumer and established the first known empire. He shifted the center of power northward and extended the domain of his activity from the Mediterranean to the Persian Gulf. However, after a time Akkad, too, began to decline. Then, in the 1700s B.C.E., Hammurabi of Babylon imposed his power on the land and Babylon became the hub of the world. But the hegemony of this particular Babylonian empire was also short-lived. Hammurabi's dynasty ended 125 years after his death, when a Cassite chieftain from the Zagros Mountains seized Babylon. The Cassite dynasty, in turn, reigned for a few hundred years, and then gave way to the rising power of the kingdom of Ashur, centered on the Tigris in northern Mesopotamia. Thus the center of civilization and power in Mesopotamia, although it later oscillated back and forth, tended in general to shift gradually northward. In any case, the land of Sumer in southern Mesopotamia never recovered its former glory or productivity.

What might have caused this northward shift? Although one must beware of simplistic generalizations, there are good reasons to believe that the decline of agriculture had much to do with it.

Ancient Mesopotamia owed its prominence almost entirely to its agricultural productivity, based on its soil and water resources and its favorable climate. It had no other resources to speak of. The soils of this extensive alluvial valley are deep and mostly quite fertile. The topography is level and the climate warm and dry, with abundant sunshine year round. The scant rainfall poses no appreciable problems of water erosion or of nutrient leaching, such as are encountered on sloping lands in more humid regions. Although wind erosion can be severe on dry soils stripped of vegetation and crushed by tillage, the main problem of crop production in this arid region is, of course, water. Fortunately, water is available in large quantities in the twin rivers: the Euphrates, and the Tigris.

The two rivers begin in the snow of the high Armenian mountains and flow down into the head of the Persian Gulf, a distance of about 1,800 kilometers for the Tigris and over 2,800 kilometers for the more circuitous Euphrates. The plain itself is an alluvial deposit of sand and silt that has been brought down by the rivers since past geological times. Historians once believed that at the time of the earliest Mesopotamian civilizations the Persian Gulf coast might have been farther northwest, thus providing the Sumerian cities with a nearby seacoast. Subsequent studies by geologists, however, have indicated that, despite the continuous deposition of sediment by the rivers, the land area in the lower valley has not increased significantly, owing to the gradual geological subsidence of this valley.

Primeval nature was represented in Mesopotamian mythology as monstrous chaos which needed to be overcome by the constant labor of people, supported by patron gods who encouraged the construction of works that would establish earthly order and regularity. Thus, the mythical feat of the hero-god Enlil, or Marduk, in slaying the goddess of chaos Tiamat and creating the world from her sundered body, reflected the work of the human Mesopotamians in reclaiming the swamps, raising their cities above the floodplains, and subduing the capricious rivers by means of levees and canals. They thus transformed a patchy land of drenched marshes and desiccated desert into a fertile expanse of fields and orchards.

The diversion of river water onto the valley land led to serious problems.[4] The first problem was silt. Early in history, the upland watersheds of the twin rivers were deforested and overgrazed. Erosion resulting from seasonal torrential rains proceeded to strip off the soil of those uplands and pour it into the streams, which in turn carried it as suspended sediment hundreds of miles southeastward. As the silt-laden flood waters (particularly of the Euphrates) wound their way toward the lower reaches of the valley, more and more of the sediment settled along the bottoms and sides of the rivers, thus raising their beds and their banks above the adjacent plain. Rivers that are elevated above their flood plains are notoriously unstable: during periodic floods they tend to overflow their banks, inundate large tracts of land, and from time to time change course abruptly. The silt also tends to settle in channels and fields, and thus to clog up the irrigation works.

The second and even more severe problem was salt. This problem, along with waterlogging, resulted primarily from the inexorable rise in the water table—a rise that, in the absence of adequate natural or artificial drainage, naturally follows the flood-irrigation of low-lying lands.

The problem is general. Elevated rivers continually cause seepage into the groundwater and thus tend to raise the water table. So do diversion canals and distribution ditches. Finally, the act of irrigation itself exacerbates the problem by causing even more seepage from the entire surface of the land. All waters used in irrigation contain dissolved salts. Irrigation without adequate drainage of river valley soils thus spurs the process of soil salinization. Since crop roots normally exclude most of the salts while extracting soil moisture, the salts tend to accumulate in the soil and, unless leached out, will in time poison the root zone.

In arid regions, natural rainfall is generally insufficient for annual leaching; irrigation must hence be applied in excess of crop water requirements so as to remove harmful salts by downward percolation beyond the root zone. Initially, for quite some years or even generations, the processes of groundwater salinization and water table rise are invisible and go unnoticed. Then, when the water table comes to within a few feet of the ground surface, a secondary process of capillary rise comes into play. The rising groundwater evaporates at the surface, precipitating the salts which are always present in the groundwater, especially in arid regions. This process infuses the topsoil with salt. As the salinization process advances, an irrigator might try to irrigate more and more in a desperate and frantic attempt to flush out the salt with fresh water, as before. But, in so doing he is merely accelerating the rise of the water table and thereby causing waterlogging as well as salinization. After each irrigation the salt reappears. It rises from below and blossoms out in mockingly beautiful floral patterns, more of it each time, until the soil is rendered sterile.

Working in tandem, therefore, silt and salt can destroy an entire region's irrigation-based agriculture.[5]

In southern Mesopotamia, the major water supply was by surface gravity flow diverted from the rivers via canals. Of the two rivers, it is the Euphrates that is more convenient to tap for the purpose of irrigating the alluvial plain. The Tigris is more unpredictable and swift, and it cuts into the alluvium to a depth of several meters over most of its course, making it difficult to divert water from it. Only along its southernmost stretch does it rise above the level of the plain. The Euphrates, because it travels a much greater distance from its sources, loses almost half of its water through evaporation and seepage in the Syrian desert, so it arrives on the southern Mesopotamian plain at a much lower speed. Because of its more sluggish pace and its greater load of silt, the Euphrates deposits more sediment and tends to run above the plain rather than cut into it, and it forms high levees. In antiquity, the river was a braided stream, dividing into several channels and switching courses periodically. Some of its channels were utilized at various periods as diversion canals.

Tapping into the Euphrates is a rather simple task, since the river bed lies above the level of the adjacent plain. All that is needed is to breach the levees and convey the water by gravity to level basins of arable ground. However, the river's discharge varies greatly from year to year and from season to season. Flow may be too meager at times and too great at others, so the problem of water shortage can be as acute as the problem of water excess. The flood stage of the Euphrates generally arrives on the plain in April, somewhat late to be of greatest benefit to winter crops. Submergence at this stage may endanger the maturing crops and wash away field and basin boundaries. Nor is the timing of the flood optimal from the standpoint of summer cropping, for which it is a bit early. Both during periods of scarcity and of surplus, problems of equitable distribution and disposal of water must have arisen among users variously located along the river and along the diversion canals: whether upstream and closer to the source, or downstream at the tail end of the system. Regulation of water flow therefore requires coordination, either voluntary or mandated.

The evident need for coordination in large riverine irrigation system led anthropologist Karl Wittfogel to postulate that irrigation-based societies, which he called hydraulic civilizations, necessarily tended to centralized control and even to despotic rule. Although simplistic, and lately much criticized, the concept contains a grain of plausibility.[6]

King Hammurabi, who ruled Mesopotamia around 1760 B.C.E. (and called himself the "obedient and god-fearing prince"), was strongly aware of the need for centralized control of the water originating in the two unpredictable rivers. In the famous Code of Hammurabi, the laws concerning water management were aimed primarily at preventing carelessness by the owner of one field that might cause flood damage to the field owned by a neighboring landholder. Documentary evidence indicates that Hammurabi also often directed his provincial governors to dig canals and dredge them regularly, as well as to build flood protection works such as earthen levees. Large-scale flooding, of course, was a constant danger in lower Mesopotamia, especially if the two rivers crested simultaneously. Ironically, the very means intended to contain flooding also increased the destructive power of a flood on the rare but terrible occasions when the rivers rose above those levees. The probability and frequency of such catastrophic events increased as the river beds rose by progressive deposition of silt brought from the denuded upper watersheds.

The sediment brought by the rivers is deposited in the channels and along their banks, and during inundations in the flooded areas as well. The pattern of deposition is such that the coarser particles settle closer to the source and form more permeable soil, whereas the finer particles are carried farther and settle downstream to form less permeable soil. The latter is, of course, more difficult to drain. Altogether, these differences in the distribution of water and silt create a patchwork landscape that strongly affects the patterns of salinity and waterlogging. The clogging of irrigation canals by silt reduces their discharge, a fact that affects the downstream users most acutely. The canals require periodic dredging, which is another task that calls for cooperation. Silt deposited in the fields raises the ground level. Although the silt adds nutrients, it increases the difficulty of bringing water to the land.

The buildup of soil salinity affects the choice of crops that can be grown. Although wheat is generally the preferred food grain, it is more salt-sensitive than barley. Cuneiform sources suggest that the proportion of barley grown in ancient southern Mesopotamia increased progressively in time, along with the spread and increase of soil salinity.[7] In more recent times, the relative areas of the two alternative crops have ranged from more than 90 percent wheat north of Baghdad to more than 90 percent barley in the southern section of the plain, which is more severely affected by salinization.

At present, the method of preventing or remedying salinization involves lowering the water table by means of lateral subsoil drainage (through open ditches or buried perforated tubes) based on gravity-flow, or by means of upward pumping through wells. In ancient times, however, practically the only method for dealing with salinization was a system of alternate-year fallowing. At the end of an irrigation season, the water table is typically within a half-meter to a meter of the surface. If the land is kept fallow for a year, it is normally invaded by phreatophytes—the native weeds capable of sending down roots to the water table and drawing water from it. These plants can lower the water table to a depth of perhaps two meters, thus lessening the capillary rise of water and the migration of water-borne salts toward the surface. Drying out the subsoil in this manner facilitates the downward leaching of salts from the soil by rainfall or subsequent irrigation.

But while the ancient practice of alternate-year fallowing may have retarded salinization, it could not prevent it.[8] At each alternate-year irrigation season, more salt was added to the subsoil and the groundwater. In the absence of sufficient natural or of artificial groundwater drainage, the fallowing system gradually lost its effectiveness. The inevitable result was progressive soil salinization. For a time, farmers may have tried to maintain productivity by laboriously scooping up and removing the badly salinized toplayer and working the deeper layers of the soil, but these layers too would become salinized in turn until eventually there could be no recourse but to abandon the land.

The civilization of Mesopotamia depended on the complex dynamic interplay between the unseen rains falling on the distant mountains, the resulting unforeseeable flooding of the rivers, and the equally mysterious upsurge of the groundwater. Life was thus seen as a contest between the fresh water from above, represented by the good god Apsu, and the poisonous brine welling up from below, represented by the evil goddess Tiamat. The story of this contest is related in *Enuma Elish,* the Babylonian Genesis. The dual origin of water is echoed in the story of the Deluge in the Biblical book of Genesis, wherein the water engulfing the earth is said to have come from two sources: "the windows of the heavens" and "the fountains of the great deep."

In a Sumerian myth, the goddess of love and procreation was envied by her sister and enemy, the goddess of death. The latter meted out her revenge stealthily, in the form of salt-laden water oozing up from below to kill the life-giving soil.

To the east of Mesopotamia, far across the deserts of southern Persia and of Baluchistan, lies the Indus River Valley. Here, another irrigation-based civilization developed in ancient times, probably under the influence of Mesopotamia. Though the Indus River civilization (sometimes called "Harappan," after the name of one of the major sites of excavation) apparently embraced an area more extensive than that of either the Sumerian or the Egyptian, much less is known about it. No written records have been discovered, hence the language of this civilization has not yet been deciphered and its history and fate cannot be determined with certainty. Some have conjectured[9] that it came to an end in a catastrophic flood of the River Indus due to tectonic disturbances. It appears more likely that this civilization, like the Sumerian, succumbed to environmental degradation. Unlike Mesopotamian cities, which were built of sun-dried mud bricks, the Indus Valley cities were built mostly of baked bricks. The firing of these bricks must have required great quantities of wood. This, combined with other uses of wood, probably resulted in widespread deforestation and denudation, which may well have been exacerbated by the grazing of cattle, goats, and sheep. The resultant erosion, and the increased silting of the river valley, may have exacerbated the periodic flooding of the cultivated land. Salinization was very probably another acute problem in the ill-drained Indus Valley then, as indeed it is even now.

*And there was famine in all lands but
in the land of Egypt there was bread.*

GENESIS 41:54

12

THE GIFTS OF THE NILE

NORTHEASTERN AFRICA is a forbidding desert, part of the great Sahara that stretches in a wide belt from the Atlantic across all of North Africa and into Southwest Asia. Within this arid and barren wasteland, incongruously, runs a fertile strip of life. It is the emerald-green valley of the Nile, a river that slithers like a creeping vine through the desert giving rise to the earth's greatest oasis. The roots of this vine draw vital water and fertile silt from distant lands, long unknown to the people of Egypt who depended on the Nile and who believed that it sprang miraculously out of the desert.

The Nile is the world's longest river, with a total length exceeding 6,000 kilometers. Spring and summer rains on the headwaters of the Blue Nile and its tributaries, which nearly always occur at the same time of year, commonly charge the Nile with fifty times its discharge during the low stage. The floods which occur in the fall of each year bring water and new alluvium to the fields of Egypt. The contrast between the fertile dark soil watered by the Nile and the barren reddish soil of the desert lying just beyond its banks is extreme and abrupt. The dark land itself is divided into two sections: The long, narrow strip running alongside the upstream section of the river is called Upper Egypt. The broad delta where the river fans out into a series of distributaries, called Lower Egypt, includes two-thirds of Egypt's arable land.

The ancient Hebrew name for Egypt is Mitsrayim (Arabic *Misr*), meaning a narrow and constricted place.[1] The ancient Greek name for Egypt was Khemia, from the word Khami signifying black earth, which was what the Egyptians themselves called their land. Our term "chemistry"[2] was derived from the name of that dark deposit of the Nile that was considered the prototype and mother lode of all material substances.

Early Greek visitors to Egypt were so intrigued by the regularity of the Nile's flooding that some believed the river had been created along with the world. It was considered too copious and unique to have had the same origin as other rivers, especially as it seemed to gush out of a vast, dry desert. The Greek geographer and historian Strabo stated of the Nile that "its rising, and its mouths . . . are amongst the most remarkable, the most wonderful, and most worthy of recording of all the peculiarities of Egypt." The Egyptians themselves believed that the Nile was divine. Hapi, their Nile god, was represented by the figure of a rather chubby man with large breasts and a clump of papyrus reeds adorning his head.

The Nile valley has nourished and sustained more than five millennia of civilization without interruption.[3] The durability of Egyptian civilization stands in sharp contrast to the fragility of its Mesopotamian counterpart. That contrast seems puzzling. The Tigris-Euphrates and the Nile are similar in that they are exotic rivers, drawing their waters from outside the arid regions into which they flow. Mesopotamia and Egypt had similar climates and raised similar crops. Both depended almost entirely on irrigated agriculture. Although civilization in Mesopotamia developed somewhat earlier than in Egypt, the two cultural centers were contemporaries for long periods of history. In fact, the two constituted the diametric great powers, the East and West of antiquity, relatively more influential in their heyday than the United States of America and the Soviet Union are today. Both civilizations had their periods of greater and lesser ascendancy. At various periods, both won and lost wars, dominated other nations, and were themselves subjugated. Moreover, both suffered periods of catastrophic floods that caused famine and periodically decimated the population.

Through all these vicissitudes, the irrigation-based civilization of Egypt survived and continued in the same location.[4] In contrast, the civilizations of Mesopotamia—Sumer, Akkad, Babylonia, and Assyria—each in turn, rose and then declined and disappeared, as the center of population and culture shifted gradually from the lower to the central to the upper parts of the Tigris-Euphrates valley.

What explains the persistence of one and the demise of the other? The answer can be found in the different soil and water regimes of the two countries. Neither the clogging by silt nor the poisoning by salt were as severe along the Nile as they were in the Tigris-Euphrates plain, so the land of Egypt could remain productive while the land of Mesopotamia suffered degradation.

In the words of Herodotus, Egypt is the gift of the Nile. There were actually two gifts: silt and water. The Delta and the narrow flood plain leading to it are deposits laid down by the mighty Nile, whose water and sediments are derived from central and eastern Africa. The silt comes mainly from the steep and rugged volcanic highlands of Ethiopia, lashed each summer by the torrential monsoonal rains rolling in from the Indian Ocean. The downpours scour the slopes, scraping off their loose mantle of mineral-rich brown soil and splashing it into the boisterous annual flood of the Blue Nile. The good fortune of Egypt, is thus derived from the misfortune of Ethiopia. Added to that silt is the humus contributed by the White Nile from its jungle and swampy sources. When in ancient times the gathering annual flood reached Egypt proper, it would overflow the river banks and deposit an amount of silt estimated to have been about one millimeter thick on the flood plain. This amount was not so excessive as to choke the irrigation canals or cover young seedlings, but it was fertile enough to add nutrients to the land and nourish its crops.

Whereas in Mesopotamia the inundation usually comes in the spring, and summer evaporation tends to make the soil saline, the Nile begins its rise in the middle of August and attains its maximum height at the beginning of October. In Egypt the inundation comes at a much more favorable time: well after the spring harvest, and after the summer heat has killed the weeds and aerated the soil.

The narrow floodplain of the Nile (except in the Delta), as well as the deep-cut nature of the riverbed, made it impractical to divert and convey water in long canals as was done in Mesopotamia. Thus, there was no widespread raising of the water table. The water table was controlled by the stage of the river, which, over most of its length, normally lies below the level of the adjacent land. When the river crested and inundated the land, the seepage naturally raised the water table. As the river receded and its water level dropped, it pulled the water table down after it. This all-important annual pulsation of the river and the associated fluctuation of the water table under a free-draining floodplain created an automatically repeating, self-flushing cycle by which the salts were leached from the irrigated land and carried away by the Nile itself.

The early farmers of Egypt, around 5,000 B.C.E., probably relied on natural irrigation by the unregulated floods to water the banks of the river. As soon as the flood withdrew, they could cast their seeds in the mud. At times, however, the flood did not last long enough to wet the soil thoroughly, and then the crops would fail and famine would ensue. So the Egyptian farmers learned to build dikes around their plots, thus creating basins in which a desired depth of water could be impounded until it soaked into the ground and wetted the soil fully enough to sustain the roots of crops throughout the growing season. The basins also retained the vital silt and prevented it from running off with the receding floodwaters. When the amount of water was excessive or untimely, the dikes could simply be breached so as to release the surplus and return it to the river.

The earliest pictorial record of artificial irrigation is the mace-head of the so-called Scorpion King[5] (circa 3,100 B.C.E.) that depicts the ceremonial cutting of an irrigation channel. The king is shown holding a large hoe, with some laborers excavating the channel and others holding a woven basket and a broom, all standing alongside the channel. A row of rectangular irrigation basins can be seen in the background.

The basis of Egypt's productivity was the nearly optimal combination of water, soil, nutrients, and organic matter, provided by a regular annual regime that was more dependable and timely than the capricious floods of Mesopotamia. It enabled Egyptian farmers to produce a surplus that fed the artisans, scribes, priests, merchants, noblemen, and—above them all—the Pharaohs who used their coercive power to order the building of self-aggrandizing monuments. Those monuments still stand today, less in testimony to the vainglorious kings who ordered them than to the diligence and organization of a society of labor rooted in the land.

Unfortunately for students of Egypt's history, the records of dead kings and noblemen are disproportionate to those relating to the ordinary life of the people. Since Egyptians wrote on sheets made of the pulp of papyrus reeds, most of their records were destroyed by humidity and time. Enough, however, has remained in papyri and wall paintings to give us a fair picture of ancient Egyptian agriculture. It appears to be remarkably similar to the agriculture practiced by the *fellahin* who live and work along the Nile today. No nation has lived so long and in such harmony and intimacy with the soil. In the words of an ancient tomb inscription: "I live, I die, I am Osiris. . . . I grow up as grain . . . the earth has concealed me. I live, I die, I am barley. I do not pass away."

The political and ecclesiastical structure of Egyptian society was strongly hierarchical. With the nearly absolute authority and presumed deity of the Pharaohs, who held sway over the cosmic order and over famine and plenty, ancient Egypt conformed perhaps more closely than Mesopotamia to Wittfogel's concept of a despotic hydraulic civilization.[6] It was the patient and arduous labor of the simple tillers of the soil that sustained the agricultural civilization of Egypt through six millennia. They persevered in the face of all hardships. Sand blown in from the

desert regularly encroached upon the cultivated land and needed
to be removed. Dikes and canals had to be rebuilt after each
inundation. And, though it was more regular than the rivers
of Mesopotamia, even the Nile could default, and the seven
lean years that Joseph predicted could indeed become dire reality.
When it did, the population suffered the consequences.

As the population of Egypt grew despite periodic setbacks,
the necessity arose to intensify production. Instead of just one
crop per year, the Nile-valley farmers could grow two, three,
or even four, given the year-round warmth and abundant sunshine
of the local climate. To do so, they needed to draw water at
will from the river or from shallow wells dug to the water
table. At first they drew water manually in buckets, which
they then carried using shoulder yokes. In time, a new technology
was invented: mechanical water-lifting devices.[7] The simplest
of these was the *shadoof*—a long wooden pole used as a lever,
with the long arm serving to raise bucketfuls of water and the
short arm counterweighted with a mass of mud. A more sophisticated device, invented some centuries later and attributed to
Archimedes, is the *tamboor,* which consists of an inclined tube
containing a tight-fitting spiral fin. Both these devices are human-
powered. The most elaborate of the ancient mechanical water-
lifting devices is the animal-powered *saqiya* waterwheel, probably
introduced to Egypt from Persia during the Persian or Greek
(Ptolemaic) occupations. All these devices are still in use today,
alongside modern motorized pumps.

The affluence of the river lands, in Egypt as in Mesopotamia,
has always attracted the hungry dwellers of the bordering deserts.
A document of the nineteenth century B.C.E. mentions some
nomads begging "to serve the Pharaoh" as "the desert was dying
of hunger." Egyptian border garrisons generally warded off such
would-be intruders. Occasionally, some were admitted. The Bible
relates that Abraham, and later Jacob and his sons, went down
to sojourn in Egypt. Egyptian texts of about 2,000 B.C.E., and
again in the thirteenth century B.C.E., state that Asiatic herdsmen
were permitted to enter Egypt "as a favor, to keep them and
their cattle alive." After the middle of the eighteenth century,

"wretched Asiatics" invaded and conquered lower Egypt. These intruders, called Hyksos—meaning "rulers of foreign lands"—were expelled two centuries later. Egypt finally lost its independence to the Greeks (who built Alexandria), and was later ruled by the Romans and a long succession of other foreigners. Yet the agricultural base remained intact—that is, until recently.

When Napoleon invaded Egypt in 1797, he brought with him an entire contingent of scholars for the purpose of studying that mystery-shrouded ancient land. Quite apart from its dubious military or political importance, this expedition made a great contribution to knowledge. It initiated the systematic study of Egyptian history, and it led, among other achievements, to the discovery of the Rosetta Stone and to the deciphering (by Jean François Champollion) of ancient Egyptian hieroglyphics. More relevant to our topic is the fact that French agricultural experts surveyed the Nile Valley and conducted a census there.[8] They reported that the population of Egypt in the early 1800s was less than 3 million. The historical evidence suggests that the population of ancient Egypt had fluctuated over the millennia from perhaps 1.5 to 2.5 million. So the population at the beginning of our nineteenth century was not much different from what it had been throughout the entire history of Egyptian civilization to that point. Today, the population of Egypt is about 55 million.

This means that 20 times as many people as in ancient times are now striving to eke out a livelihood on the same soil and the same water-resource base. In ancient times, Egypt was able to export food and helped to sustain the Roman Empire. Today, Egypt imports more than half of the food it needs. Worse yet, its soil—legendary for its durability and productivity—is now progressively deteriorating.

*The land ye go over to possess is a land
of hills and valleys, that drinketh water
as the rain of heaven cometh down.*

DEUTERONOMY 11:11

13

HUSBANDRY OF THE
RAIN-FED UPLANDS

THE UPLAND REGIONS surrounding the Mediterranean provide a telling example of how civilized man has tended to destroy his environment.[1] Many of the areas that once supported a thriving agriculture are now largely unproductive and sparsely populated. The face of the land itself is a more eloquent and revealing document than all the written records.[2]

The historical pattern is similar throughout: enterprising farmers began to clear the hillslopes of their natural vegetative cover of trees, shrubs, and herbs in order to raise crops on the fertile soil that had been formed by nature over the previous millennia. As more and more of the land was denuded and brought under cultivation, erosion stripped away the topsoil. In addition, continuous cropping hastened the leaching of nutrients and the decomposition of the soil's organic matter without replacement. Measures of soil and water conservation were insufficient or inconsistent. Overgrazing, while trampling and loosening the soil, further destroyed the vegetative cover and prevented its regeneration. The land's initially high productivity declined, and the societies dependent on it declined accordingly.

Some nations prolonged or even enhanced their prosperity

95

temporarily by conquering the land of others and exacting tribute. That may well have been a primary reason why the Phoenicians, Greeks, Carthagenians, and Romans, each in turn, were compelled at some point in their history to venture away from their own land and to establish far-flung colonies and empires in order to control larger and larger areas (*lebensraum* being the equivalent twentieth-century term). However, even the most aggressive and organized of these empires eventually failed to maintain their ascendancy because they could not go on indefinitely conquering ever more land to satisfy their ever growing requirements. Their energies were sapped by the incessant need to quell the repeated rebellions of resentful subjects.

To understand the fate of the lands and peoples of the Mediterranean basin, we must examine the soil-water-climate conditions governing the region's agriculture. The Mediterranean climate is characterized by an annual cycle of a rainy winter and a totally dry summer. The annual rainfall ranges from well under 200 millimeters in the arid southern subregions to over 1,000 mm in the semihumid uplands fringing the Mediterranean littoral along the north. Although the rainfall in the uplands is not generally abundant, and not uniformly or ideally distributed in time and space, it is generally sufficient to support a rich natural community of trees, shrubs, and herbs. When that vegetation was initially cleared, generally by fire, the combination of rainfall and fertile loamy soil could support the growth of many crops.

Winter rainfall, in principle, is more effective than a comparable amount of summer rainfall, because of the lower evaporative demand during the cooler season. Therefore, a rainfall of 400–500 mm here is equivalent to perhaps 600–700 mm in a region of summer rainfall—such as Arizona, for example—and could be sufficient to raise a perennial crop of fruit trees, or an annual crop of either a winter or a summer grain. (The growing of a summer crop would require fallowing the land in winter in order to collect and conserve soil moisture for crop use during the following rainless summer season.) The marginal threshold rainfall for winter cropping is about 300 mm; less rain than

that would generally result in crop failure, and more would
give a proportionately higher yield.

The effectiveness of rainfall in sustaining crops depends on
the presence of a receptive and retentive soil. The soil must be
able to absorb the rain rather than shed it, and it must be able
to store it in the root zone. The typical soil of the Mediterranean
uplands is called *terra rossa*. It is a loam or a clay-loam, formed
on hard limestone, and it is noted for its red color and favorable
structure. Another prevalent soil, called *rendzina,* is a darkish
or grayish loam formed on soft limestone and on chalky or
marly bedrock. Both types of soil are relatively receptive to
rain. However, the amount of water any soil can retain depends
on its depth. And herein lies the problem. These residual upland
soils tend to be rather shallow, covering the hillsides to a depth
that seldom exceeds one meter, and on steep slopes is much
less than that. Only on flat plateaus and particularly in intermontane valleys, where sediment is deposited by gravity and water,
does the soil attain much greater depths.

When rain falls on sloping land, part of it infiltrates and
part runs off, in varying proportions.[3] A soil that is deep, well-
structured, and covered by protective vegetation and a mulch
of plant residues—that is, soil in its natural state—will normally
absorb 95 percent or more of the rainfall. On the other hand,
a soil that is denuded of vegetative cover and deprived of a
surface mulch may absorb less than 80 percent of the rain, and
in extreme cases less than 50 percent, especially if the rainstorms
are intense, the soil is shallow, and its surface has been compacted
by tillage, by trampling, or by the raindrops themselves. The
effect is to reduce the amount of available moisture in the soil
and thus make it a more arid environment for crops. As more
runoff is induced, accelerated erosion ensues. The water trickling
off the slopes causes frequent flooding and a rise of the water
table in the bottomlands. That problem is exacerbated by the
sediment which clogs the natural drainage outlets. Consequently,
malaria-infested marshes often formed in the coastal and intermontane valleys of the Mediterranean littoral.[4]

Erosion on hillslopes not only deprives the soil of the nutrients and humus naturally concentrated in the topsoil, but also reduces the thickness of the rooting zone and its capacity to absorb and store moisture for crop needs. In extreme cases of erosion, all that remains of the original continuous mantle of soil covering the slopes is a series of shallow pockets of soil among exposed outcroppings of bedrock. The greater the portion of ground with exposed bedrock, the less the area's receptivity to rain. So the fraction of rainfall that is lost by surface runoff increases correspondingly. A negative feedback cycle is thereby generated: less vegetation causes more erosion, hence less infiltration and still more runoff and erosion, in a self-accelerating process of degradation.

This insidious process can be especially severe in the Mediterranean region, where the rains do not appear as gentle showers or drizzles, as they generally do in Northwestern Europe, but as violent and quite erosive squalls. Ironically, the beneficent rain that is so vitally needed and eagerly awaited by farmers can become a voracious monster, gnawing at the soil and wearing away the land.

The real problem of rainfed agriculture on sloping ground is how to control the erosive power of rainstorms, and how to promote the penetration of water into the soil rather than let it escape as surface runoff. Unless this problem is solved, there can be no sustainable agriculture on the rainfed uplands, just as there can be no sustainable agriculture in the irrigated river valleys unless the twin problems of salinization and waterlogging are solved. So rainfed farming in the Mediterranean region, which began in the flat intermontane valleys where erosion was not a serious problem, perforce had to devise specialized methods of soil conservation as it gradually expanded—under pressure of increasing population—onto the uplands.

The story of the Israelite settlement in the land of Canaan is particularly instructive.[5] Although it is a small country, its climate is highly variable, from semihumid in the north to extremely arid in the south, and it has a diversity of ecological conditions. In the northeast is a beautiful fresh-water lake, the Sea of Galilee, into which flows the swift and exuberant stream of the upper Jordan. Out of that lake, the lower Jordan descends on its tortuous and ill-fated journey south through the searing Rift Valley, only to die a tired death in the thick steamy brine of the land-locked Dead Sea, in the desolate Valley of Sodom and Gomorrah. The northern and central parts of the country, however, are relatively humid, with high hills, intermontane valleys, and a coastal plain. The hills of the central part of the country grade gently westward—facing into the path of rain-bearing winds—toward the sea, but fall steeply toward the arid Rift Valley in the east. The southern part of the country is the Negev desert, which blends into the Sinai desert.

To the destitute and desperate nomads coming out of that desert at the beginning of the Iron Age, circa 1,200 B.C.E., the Promised Land must have seemed like a veritable paradise. However, it was a land already well-populated, in which the most desirable areas were filled to capacity. Finding the valleys occupied, the Israelites had no recourse but to settle in the central mountain ranges of Judea, Samaria, and Galilee. Here they faced new and unfamiliar challenges: a variable and some-times capricious pattern of rainfall; few and meager sources of water for domestic needs, and practically none for irrigation; shallow, stony, and erodible soils; a rugged terrain with practically no flat land; and a thicket of oaks, pines, and dense shrubs that the settlers needed to clear away. So they had to learn the ways of water and soil conservation: how to hew out and plaster cisterns so as to collect and store rainwater for the dry season, how to tap groundwater and springs,[6] and—especially—how to carve out arable fields on steep slopes. They did this by collecting stones off the ground and using them to construct rock walls on the contours, thus dividing the slope into a series of terraces. Furthermore, to restore soil fertility they allowed the land to rest for one year out of seven, a sort of sabbatical known as *shmitah*.[7]

The back-breaking work that was required is described in numerous passages of the Bible and the later Hebrew writings. Particularly poignant is the poetic outcry in the Book of Isaiah[8]:

Let me sing of my well-beloved, a song of my beloved concerning his vineyard. My well-beloved had a vineyard in a very fruitful hill. And he digged it, and cleared it of stones, and planted it with the choicest vine, and built a tower in the midst of it, and also hewed out a vat therein. And he hoped that it should bring forth good grapes, but it brought forth sour grapes. . . . And now come, I will tell you what I will do to my vineyard: I will take away the hedge thereof, and it shall be eaten up; I will break down the fence thereof, and it shall lay waste: it shall not be pruned nor hoed, but there shall come up briers and thorns, and I will command the clouds that they rain no rain upon it.

The technique of constructing walled terraces probably began in the latter part of the second millennium B.C.E. This technique gradually became an important feature of the Iron Age expansion of agricultural settlement in the hill country. The aims of terracing were simple: to transform sloping ground into a series of nearly horizontal arable plots with adequate control of water and minimal erosion. In the process, however, the practice of terracing stamped the permanent imprint of man on the landscape. It transformed the natural slopes, altered the patterns of natural drainage and erosion, changed the profile and development of soils, and produced culturally controlled patterns of flow and sedimentation.

If the initial construction of terraces was an arduous task, so was the subsequent requirement of continuous maintenance. Each terrace consisted of a carefully built rock wall with a horizontal crest, serving to support and retain the upslope soil. The soil could then be levelled to form a flat field-plot of sufficient size and depth to allow cultivation and crop husbandry. Obviously, such an undertaking could only be feasible where there was secure land tenure and communal cooperation.

Ironically, the terraces which protected the soil also increased its vulnerability to erosion. Such terraces are effective in conserving both soil and water only as long as they are perfectly maintained; if neglected, they increase the hazard they were designed to abate. The moment a terrace wall is breached, whether by the spontaneous collapse of stones under the weight of the soil, or by grazing goats trampling over and dislodging the stones, the exposed vertical mass of soil is certain to slump

and erode even more rapidly than before the terraces were built.
The farmers of ancient Israel undoubtedly cared for their terraced
fields as long as they were allowed to live and work peacefully
on their land.

Unfortunately, wars would intervene all too frequently. Invading armies would plunder the villages and ravage the land.
At times they would even send the entire population into exile,
as the Assyrians did to the northern Israelites in 722 B.C.E.,
the Babylonians to the Judeans in 587 B.C.E., and the Romans
again to the Judeans in the first and second centuries C.E. In
the wake of such disasters, the sedentary population would be
replaced by bands of herding nomads, and their sharp-hoofed
goats would trample over the land indiscriminantly, devouring
the vegetation and destroying the terraces. Indeed, the hills of
Israel and Lebanon are full of relics of ancient terraces which
had fallen into such a state of disrepair that the soil they once
held has long since been eroded down to bare bedrock. Some
of the terraces, however, have somehow—against all odds—remained intact through the ages and are still cultivated and productive, proving that the system is indeed sustainable if properly
maintained. Sadly, many of these terraced plots are at present
considered too small and too irregularly shaped to allow efficient
farming with modern machinery.

The Near Eastern goat, incidentally, has been vilified and
condemned too harshly by some presumed experts as an inveterate
and incorrigible overgrazer, a denuder of vegetation and promoter
of soil erosion, a cause of desertification, and altogether as a
menace to civilization. The truth is that overgrazing is caused
by the people who put too many animals for too long on an
area of rangeland incapable of sustaining them.[9] With proper
grazing management, the goat is no special menace, and no
worse than its cousin the sheep. With excessive grazing, however,
and given no choice, goats will subsist on anything they can
find, including thorny and dry brush, and will survive long
after the sheep have died of starvation. The goat is thus a remarkably hardy animal, a spirited and whimsical individualist who
fully justifies the application of its name, *capra,* to the adjective

"capricious." It is through no fault of its own that it became an agent of great destruction throughout the Near East.

The history of land husbandry in Lebanon is as instructive as the case of Israel.[10] Lebanon was the land of the Phoenicians, who settled along the coast sometime during the second millennium B.C.E. During the period from 1,000 to 500 B.C.E. they became the Mediterranean world's foremost navigators and traders. As conduits of culture, as well as of commercial goods, they transmitted the Semitic alphabet to the Greeks. Lebanon received enough rain to produce good yields of grain, grapes, olives, and many other crops, but the amount of tillable land along the coast was limited by mountain ranges. These mountains were covered by dense stands of the famed cedars of Lebanon. The wood was used for construction of buildings and of ships, and became a prized commodity of international trade. So the Phoenicians carried out systematic logging of the cedars. The wood was sold to treeless Mesopotamia and Egypt. The Biblical Book of Kings reports that King Solomon, with the agreement of the Phoenician King Hiram, sent scores of thousands of laborers to Lebanon to bring timber to Jerusalem for the construction of his temple and palaces.

With prosperity came an increase in the population of Phoenicia. Following the deforestation, cultivated fields began to creep up the hillslopes, and erosion ensued. By the ninth century B.C.E. the Phoenicians found that their agriculture and their extensive commerce—based on exporting lumber and various industrial products, including glass and dyes—were inadequate to support their growing population. They then embarked on colonization of other lands, including Carthage on the coast of North Africa, and other colonies in Sardinia, Sicily, and Spain. These colonies supplied food for the home country and accepted its exports. But as Phoenicia's soil gradually deteriorated, so did the strength of the nation. Finally, the Phoenicians were defeated by the Greeks, whose King Alexander destroyed the city of Tyre in 332 B.C.E. Carthage established itself on new lands, and continued on its own for two centuries more, but succumbed to the Romans in 146 B.C.E.

The Greeks who conquered the Phoenicians, and then the Romans who conquered the Greeks, eventually suffered the same fate. The sordid story of soil erosion repeated itself in each of the countries around the Mediterranean where farming was practiced on the hillsides.

Greece is a hilly country, with few extensive valleys. Its hillsides, originally forested, were blessed with a fertile soil that was, however, rather shallow and vulnerable to erosion. Greek civilization greatly affected the environment within which it operated. It transformed a land once densely vegetated into a terrain of naked rocks. This severe alteration of the landscape must have contributed to the decline of Greece as a center of culture and power. The Greeks cut their trees for fuel. They turned the wood into charcoal, which they used for firing pottery and bricks and for reducing mined ores. They also used wood to construct roofs, doors, and furniture; carts and chariots; and especially ships. Greek shepherds did additional damage to the forests by deliberately setting fires to eradicate the woody vegetation and encourage the growth of grass, which they then overgrazed.

Greek farmers added to the damage by growing wheat, barley, olives, figs, grapes, and various other fruits and vegetables on sloping ground with insufficient protection of the soil. They plowed for grain in the autumn with oxen, using a simple plow with an iron share, planting wheat in the October or November rains and harvesting it with a sickle in May or June. Once the land was bare of its vegetative cover and the soil was loosened, the torrential rains of the Mediterranean (occurring during the autumn, winter, and spring seasons) began to wash away the topsoil. Still more destruction was wrought by warfare, in the course of which each army would deliberately devastate the farms and fields, and cut down the trees, in the land of its enemies. The ensuing erosion was described quite vividly in the *Iliad:* "Many a hillside do the torrents furrow deeply, and down to the dark sea they rush headlong from the mountains with a mighty roar, and the tilled fields of men are wasted." Thus was destroyed the ground that might have grown trees again.

The sediment raked off that ground was laid at the mouths of the streams, clogging their outlets and causing the formation of swamps, breeding swarms of malaria-infested mosquitoes.

Upland grazing and then farming in Greece probably began around the middle of the second millenium, and was greatly intensified during the Hellenic period, from 800 B.C.E. As the soil eroded, the Greeks had to shift from food grains to commercial crops of grapes and olives, which could be grown on thinner soils. The fruit was processed into wine and oil for export, but the increasing reliance on trade made the Greek economy still more vulnerable. Erosion continued unabated in any case. The soil that was stripped off the slopes and carried into the streams clogged their outlets and formed marshes. By the time the Macedonians assumed their hegemony over Greece in 338 B.C.E., the land had already deteriorated markedly. Of necessity, the Greeks then committed themselves to pursuing the career of conquests they had actually begun earlier. In the course of this venture, while their own home country continued to decline, they transplanted their Hellenic culture and power to such distant centers as Antioch on the Mediterranean coast of Syria, Seleucia-on-the-Tigris, and especially Alexandria, which they established as a seaport to transport Egyptian grain. During the later Hellenistic period, efforts were made to expand the agricultural land of Greece by draining marshlands, as well as the Boetian Lake Copais. Yet the erosion process continued.

Plato, in one of his dialogues, has Critias proclaim:

> What now remains of the formerly rich land is like the skeleton of a sick man, with all the fat and soft earth having wasted away and only the bare framework remaining. Formerly, many of the mountains were arable. The plains that were full of rich soil are now marshes. Hills that were once covered with forests and produced abundant pasture now produce only food for bees. Once the land was enriched by yearly rains, which were not lost, as they are now, by flowing from the bare land into the sea. The soil was deep, it absorbed and kept the water in the loamy soil, and the water that soaked into the hills fed springs and running streams everywhere. Now the abandoned shrines at spots where formerly there were springs attest that our description of the land is true.

Two centuries earlier, Solon had already advocated discontinuing grain cultivation on the sloping lands in Attica, and recommended planting olives and grapes instead. His advice was echoed in the fourth century B.C.E. by Theophrastus in his *Cause of Plants*. However, neither man's advice addressed the root cause of the problem, which was not the choice of a crop as such but the process of erosion and the failure of the ancient Greeks to control it. It is a tragedy of history that these remarkable people did not apply their outstanding ingenuity and diligence more specifically to the vital task of conserving and managing their soil and water resources.

Rome, no less than the Greek cities, initially depended on local agriculture. During the early phase of its history Rome's economy was based on utilizing the fertile soils, lush vegetation, and benign climate of the central part of the Apennine Peninsula. However, like Greece, the land of the Romans is hilly. And though they too worshiped nature deities and professed to revere the earth, calling it *mater terra,* the Romans were strictly utilitarian in their attitude toward the land, as toward most other aspects of life. Their fundamental belief was that the earth and everything on or in it was meant for human use, for their use. In the short run, this attitude seemed practical and led the Romans from success to success, an experience that bolstered their sense of superiority. Their hubris-driven approach to land utilization impelled them to divide the territory under their control into geometrically drawn squares, regardless of topography, and to organize their farming methods accordingly.

Soon the Romans began to notice that their soil was becoming less fertile than it had been. Far from attributing this decline to mismanagement, they adopted the belief that the world as a whole was growing old and feeble, and that only vigorous human intervention could maintain it. This belief was expressed by some of their most prominent thinkers, including Virgil, Ovid, and Seneca. Cicero, for one, heaped lavish praise upon human cleverness at manipulating nature in the areas of agriculture, mining, construction, water supplies, and navigation. In his words, "By means of our hands we endeavor to create as it were a second world within the world of nature." Seneca, too, held that human efforts make the world more beautiful and useful. Pliny the Elder, however, and the wise agriculturist Columella, ascribed the declining fertility of the earth to poor husbandry.

Roman farmers were aware, of course, of several methods for enhancing soil fertility, including fallowing, manuring, and terracing, but they applied these methods only sporadically. The trend toward consolidation of farmland and the ownership of larger and larger estates (called *latifundia*) by wealthy absentee owners who employed slaves or impoverished peasants were not conducive to careful, and caring, husbandry. Eventually, sections of played-out land had to be abandoned. Such tracts were called *agri deserti* (deserted fields).

Seeing what was happening, Lucretius lamented that "forests are receding higher up the mountains, yielding the ground to agriculture." The result was a greatly accelerated rate of erosion. The typical upland soils of the Mediterranean countries, formed on limestone, are shallow even in their virgin state, and when eroded they are often reduced to mere pockets amid the protruding outcroppings of bedrock. Hillsides bereft of soil cover absorb less rain, so the natural rain-fed mountain springs tend to dry up. At the same time, runoff increases, so the seasonal flooding and silting of lowlands are intensified. The latter effect muddied aqueducts, clogged estuaries, created malarial marshes, and silted up harbors such as Paestum, Ostia, and Ravenna.

The Romans' mistreatment of nature was carried considerably beyond the environs of their own land. Everywhere they established their dominion, they repeated the same pattern. Forest clearing took place extensively, as did overcultivation and over-grazing of land to satisfy the avaricious demands of a bloated center of power. Especially affected were the regions of North Africa, which the Romans wrested from the Carthagenians in the wake of the Punic Wars.

In the middle of the third century A.D., the future St. Cyprian, then Bishop of Carthage, wrote to the Roman proconsul of that region: "The world has grown old and has not retained its former vigor. It bears witness to its own decline. . . . The husbandman

is failing in his field. . . . Springs which once flowed profusely
now provide only a trickle." Being a Bishop, Cyprian attributed
the decline to lack of faith.

About thirty kilometers northwest of Carthage is the site
of the ancient city of Utica. Established by the Phoenicians as
a seaport at the mouth of the Bagradas River, it gradually lost
its access to the sea because of sediment that was eroded off
the hills and then transported and deposited by the river. The
remains of the city now lie under about ten meters of silt, at a
site that is some seven kilometers from the coast. As a result
of this accumulation of sediment, the river has long since changed
its course. The fate of Utica is typical of what has befallen the
other magnificent cities established by the Romans in North
Africa.

It would be simplistic to state categorically that the fall of
the Roman Empire was due to abuse of the environment. There
were undoubtedly many other reasons—political, social, military,
medical, perhaps even climatic. Suffice it to say that environmental degradation must have been an important contributing factor.
Ironically, the very achievements which most impress so many
observers to this day, and of which the Romans themselves
were most proud (namely, the grandeur and scale of their works),
were the most destructive of the natural environment and hence
most likely to have hastened the decline of their empire.

The works of the Romans can be taken to illustrate the
difference between technology and true science. The former is
directed toward achieving specific utilitarian goals, whereas the
latter aims at acquiring a fundamental understanding of the
processes and relationships operating in the natural world, and
is therefore an open-ended quest for knowledge and insight.
The Romans excelled in "practical" technology, but had little
patience for the pursuit of knowledge for its own sake, for science.
Technology that is not based on understanding its own consequences, and is not re-examined continuously in the light of
growing and deepening science, is ever in danger of becoming
self-defeating.

*The desert shall rejoice and blossom as
the rose. In the habitation of jackals,
herds shall lie down amidst lush grass.*

ISAIAH 35:1,7

14

THE DESERT REJOICED

IN ANTIQUITY the desert was regarded as a world unto itself,
an extraterritorial realm separate from and additional to the
other two known realms: the seas and the habitable lands.
Residents of the latter viewed the strange people of the desert
with fear and hostility, perceiving them to be a threat to civilization—as indeed they often were. The desert itself was held in
awe as a place of terror, a largely useless and dangerous domain.
One ventured into its mysterious vastness only at great risk.
The desert's forbidding character has also been a challenge: a
defiance of civilized man's self-proclaimed mastery of the earth;
a barrier to human expansion, to progress, to economic development; a fortress holding out against colonization and civilization.

But just what do we mean by the term desert? In what
sense does it differ from what is commonly called an arid zone?
Aridity in general is an imbalance between the demand for water
and its supply, the supply being too scarce to meet the demand.
Obviously, there can be different degrees of aridity. Such are
the vagaries of climate that even so-called humid regions can
experience occasional drought and even prolonged dry spells,
though a humid region, by definition, is one in which annual

precipitation is sufficient to sustain crop plants, and at times
may even be excessive. A semiarid region is one in which precipitation is sufficient in most seasons but in which droughts occur
frequently enough to make rainfed farming a somewhat hazardous
venture. An arid zone is one in which rainfed farming is marginal—successful in some years, but so frequently unsuccessful
as to make rainfed farming a highly insecure venture. While
in arid zones the mean annual rainfall may be just sufficient,
frequent droughts—perhaps in one year out of three, occasionally
occurring in clusters—can cause crop failures and famines.

In arid regions, ironically, the crop plants' requirements
for water are greatest, whereas the supplies by natural precipitation are the least, so that from the outset the scales are weighted
heavily against agriculture. The imbalance must be rectified by
augmentation of water supply by irrigation whenever possible,
and by strict water conservation at all times. Despite the ever-
present hazard of drought, farming populations can and do exist
there, however precarious their economy. Extensive grazing—
in addition to regular cultivation and occasionally in preference
to it—becomes a major form of land use in such areas.

The situation is basically different in real desert areas, which
can be thought of as extremely arid. Here even the precipitation
in an average year, let alone in a drought, is definitely insufficient
to sustain agricultural crops, so regular rainfed farming is impossible. Hence the biblical definition of the desert as "the land
unsown."[1] Even extensive grazing is marginal, and often submarginal. For humans to subsist in the desert without having to
import most of their vital requirements, they must devise ingenious schemes to obtain supplementary supplies of water, either
by wresting the precious fluid from underground aquifers, if
available, or by collecting it off the slopes of barren ground
during brief episodes of rainfall, or by conveying it from another
region. Only with such measures does agriculture become possible, and then only on a fraction of the land area.

Despite all the problems, the desert is not without promise.
Although the term desert is derived from the Latin word for
abandoned or deserted, not all deserts are totally useless waste-

lands. In fact, some deserts were settled by extraordinarily diligent and ingenious people, who proved that civilization can be established even in the most difficult circumstances. Evidence of such civilizations can be found in the American Southwest, in North Africa, in Arabia, in Jordan, and, notably, in the Negev Desert of southern Israel.

I first visited the Negev and the contiguous Sinai as a youth, and was immediately captivated by the vast expanse and grandeur of the land, and by its stillness. It occurred to me that this overwhelming stillness may have been the reason why the momentous revelation of the unity of God and the universe first occurred to mortal humans in the desert, and in this of all deserts. But then I discovered that this awesome silence had been broken many times by the noise of human habitation. The Negev had once been inhabited, and it hid within its recesses the remains of no fewer than six fabled cities.

In the original Hebrew, the name Negev denotes dryness. As deserts go, it is rather small, constituting only a minuscule part of the great desert belt of North Africa and Southwest Asia. Being on the fringe of this desert belt, much of the Negev is not an extremely dry desert.[2] The mean annual rainfall decreases from 200 millimeters in the northwest to about 25 mm in the far south, and is confined to the winter months, November to April. The distribution of rainfall within the rainy season is highly irregular, and the total seasonal amount fluctuates widely from year to year.

The Negev's historical importance derives from its geographical position as a narrow land bridge connecting Asia and Europe on the one hand, with Africa (Egypt) on the other. Hence it has always served as a crossroads of trade and traffic between the continents. The advantages of controlling the region, however, were frequently offset by the disadvantages. The same routes that made trade possible and opened up cultivable areas to civilized settlement in times of peace were the ones followed by invading armies in times of war. Moreover, neighboring desert nomads were always ready to plunder the settled land and its inhabitants. Thus, to the difficulties posed by the paucity of water, the erodible soil, and the fragile vegetation, was added the require-

ment of constant vigilance against the danger of encroachment
by hostile forces.

The long procession of civilized human history in the Negev begins, as far as can be discerned, during the Chalcolithic Age, the intermediate stage between the Late Stone Age and the Early Bronze Age (that is, toward the end of the fourth millennium B.C.E.) After a lapse, a new civilization arose there in the Middle Bronze Age, between the twenty-first and nineteenth centuries B.C.E. Then the Israelites arrived, starting at the end of the Late Bronze Age (circa 1,200 B.C.E.), and continuing into the Iron Age. King Solomon and his Judean successors—Asa, Jehoshaphat, and Uzziah—established villages, fortresses, and trade routes in the deserts of Judea and the Negev, and linked their kingdom with the copper mines of the Arava Valley and with the seaport of Elath on the Red Sea. Concerning King Uzziah, the Second Book of Chronicles mentions as a major achievement that he "built towers in the wilderness and hewed out many cisterns."

After the destruction of the Kingdom of Judah by the Babylonians, a new nation took possession of the Negev and built a magnificent civilization there, the achievements of which excite the imagination and admiration of visitors to the region to this day. The new masters of the Negev were the Nabateans,[3] people who began as nomadic traders and in time became superb architects and engineers, as well as expert hydrologists and diligent cultivators. The Nabatean domain lay astride the important ancient trade routes between Arabia in the south and Syria in the north, and between the Orient, including India, and the Mediterranean world. These were the routes along which camel caravans transported spices and silks, ivory and incense, frankincense and myrrh and medicinal herbs—commodities as prized in antiquity as are perfumes, cosmetics, and drugs today. Spices were more highly prized then than now, not so much because our taste in food has become blander, but because canning, refrigeration, and other means of food preservation that we take for granted were then unknown, and food could quickly become inedible without a heavy dosage of spice.

Caravans passing through the desert needed stopping places

where they might rest and obtain water and provisions. To secure and supply their trade, the Nabateans therefore had to establish and maintain regularly spaced bases along their main routes, at important crossroads with secure sources of water. These bases gradually grew into permanent, self-supporting villages and eventually into cities, and the Negev became more densely populated than ever before. Although the Nabateans' capital, the fabled red city of Petra, was built in the Edomean mountains (in southern Jordan), their population was centered in the Negev, where they built six major cities and numerous smaller villages. To maintain a population of tens of thousands, the Nabateans perforce had to develop agriculture in order to ensure a livelihood for their people. In this task they were undoubtedly aided by the example of their predecessors. But the Nabateans excelled all previous efforts.

The trade route monopoly enjoyed by the Nabateans ended some time in the first century of the Christian era, when the Romans discovered that the seasonal monsoon winds enabled them to sail through the Red Sea to India and back. They were thus able to trade directly for the coveted spices and aromatics. Soon afterwards, the Nabatean settlements had to face their greatest test of survival. Their lucrative role as caravan stations having been usurped, these settlements in the Negev had to become self-sustaining or disappear. The Nabatean King Rabel II (70–106 c.e.) is described on coins and in inscriptions of his day as "he who brought life and deliverance to his people." He very probably earned this distinction by emphasizing the improvement of desert farming practices, by which alone the Nabatean people could thrive. The same populace continued even after the Romans annexed the region and made it a frontier province. After the division of the Roman Empire and the establishment of Byzantium, the entire eastern realm of the empire enjoyed a period of stability and prosperity. The Negev became still more densely populated, and the technical achievements of the era surpassed even those of the Nabateans when they were independent.

The eclipse of the Byzantine golden age in the Negev came

very abruptly in the seventh century. Following the Moslem conquest in 636 C.E., the disruption of the old order and its links to the Mediterranean world caused the population to dwindle. Desert nomads took over and ushered in a long period of retrogression and poverty. Where thousands once prospered, a few hundred now eked out a bare subsistence. Magnificent monuments were pried apart, or crumbled gradually into haphazard heaps of stone. Great cisterns were choked by dust, and strongly built dikes were loosened by time and left unrepaired. Complete farm systems that needed only to be maintained were left untended and allowed to disintegrate. Overgrazing the dry stream beds caused erosion, so that the formerly wide bottomlands irrigated by water-spreading methods became narrow, gouged-out gullies. Terraces once green with crops were left high and dry while torrential floods rushed uncontrolled through breached dikes and scoured the creeks. Thus, the best efforts and experience of generations of diligent people were wasted by neglect and abuse. The casual visitor to the Negev finds it difficult to understand how the ancients could have developed so grand a civilization in the midst of such barrenness. Only a careful study of their techniques can reveal the answer.

Permanent rivers are totally absent in the Negev, and even springs or proper locations for digging shallow wells are few and far between. Hence the major source of water for humans and animals could only be the collection of surface runoff obtained from sloping ground during winter rains,[4] a task that has been called "water harvesting." The ability to collect and store potable water from runoff was the first imperative of desert settlement. This was done by means of cisterns, which are artificially constructed reservoirs filled by directed surface flows during each infrequent rainstorm. The early cisterns were undoubtedly crude and inefficient. Building efficient cisterns became possible only with the advent of watertight plaster, the recognition of suitable rock formations, and the proper construction of channels to collect and divert overland flow. Cisterns, incidentally, were a time-honored means of supplying water to humans and livestock throughout the Near East. The city of Jerusalem, for instance, was sustained for many centuries by cisterns hewn in the bedrock and fed by runoff from roofs, courtyards, and streets. So, although cisterns were not unique to the Negev, they were absolutely essential there.

Where cisterns could be located along the rim of a natural watercourse, they were filled by flash floods rather than with runoff collected directly from the slopes through constructed channels. However, most cisterns in the Negev were built on hillsides and depended on the direct collection of runoff. Many hundreds of such cisterns were built in the Negev, and they are clearly discernible landmarks even today. A typical cistern resembles a giant necklace, with the glistening white pile of excavated rock hanging as a pendant from the two collection channels which ring the hill and curve down its sides from opposite directions. To the thirsty ancient traveler, to whom these cisterns beckoned from afar, no sight could be more gladdening.

Runoff water was also used for irrigating crops. The runoff from winter rains falling on adjacent slopes was gathered and directed to bottomland fields for periodic soakings, to accumulate and store sufficient moisture in the soil to produce crops. Although the Negev's average winter rainfall is only about 100 mm, the runoff farmers were able to gather and concentrate sufficient runoff from the barren slopes to develop intensive agriculture in the depressions and bottomlands, which constituted only some 5 percent of the total area in the Northern Negev Highlands subregion.

This ingenious type of desert agriculture has been called runoff farming. Whereas farmers in more humid regions aim to have the soil absorb all the rain where it falls, thus preventing runoff, the desert farmers of old worked on the opposite principle. Their aim was to prevent the rain from penetrating the soil on the slopes, producing the maximum possible runoff. They then collected this runoff from a large area of slopes and directed it to a relatively small cultivated area in the bottomlands.

The cultivated area was usually divided into small field plots, which had to be leveled and terraced to ensure the efficient

spreading of water as well as both soil and water conservation.
The oldest version of runoff farming probably consisted of terracing the small creek beds which collected the runoff naturally.
This terracing transformed the entire length of each creek into
a continuous stairway, with stairs perhaps 10–20meters wide
and 20–50 centimeters high. The terrace walls were designed
to spread the flood and to prevent erosion. The slowed-down
cascade from one terrace down onto the next could thus irrigate
the field plots sufficiently for a crop to be grown. Distinct groups
or series of terraced plots, having definable catchment areas and
surrounded by stone walls, formed integral farming units of
perhaps several hectares of cultivated land. The remains of hundreds of such farm units are spread throughout the Negev highlands, most commonly around the principal ancient towns.[5]

Detailed observation of ancient runoff farm units reveals that
each unit was served by a particular and well-delineated portion
of the watershed. An elaborate system of conduits was constructed
to collect runoff from specific sections of the adjacent slopes,
not merely for each farm or set of fields, but indeed for each
terraced field within the farm. The complete farm unit thus
comprised both the slope catchment (the runoff-contributing
area) and the bottomland fields (the runoff-receiving area). Fields
could be made productive only if associated with a catchment
from slopes, since the meager rainfall alone was far from sufficient
for any crop. The larger the catchment, the greater the water
supply one could expect and the corresponding area that could
be irrigated. Clearly defined catchment areas, allocated to serve
particular farm units, constituted "water rights."

Typical farm units consisting of 0.5 to 5 hectares were associated with 10 to 150 hectares of sloping watershed. The ratio
of runoff-contributing catchment to runoff-receiving cropland
varied from 20 : 1 to 30 : 1. If each hectare of sloping land contributed only 10 percent of its annual rainfall of 100 mm, then
the receiving cropland would have gotten approximately 25 ×
10 mm = 250 mm. Added to its own reception of the annual
100 mm of rainfall, the plot would thus have received a total
of 350 mm, just enough to produce a crop. If, however, the

runoff yield were 20 percent of annual rainfall, the amount of water received by the field could equal 500 plus 100 for a total of 600 mm, an amount equivalent to the rainfall of the relatively humid Mediterranean habitats along the coasts of Israel and Lebanon.

The fraction of runoff yielded by the watershed varied, of course, from year to year, and even more so from rainstorm to rainstorm. Gentle showers contributed practically no runoff, whereas intense squalls might yield 30 percent or more of their rain.[6] So, even with all the alertness, ingenuity, skill, and diligence they could muster, the runoff farmers of the Negev operated a risky business and had to face new uncertainties each season. It is all the more remarkable, therefore, that they were able to cope with all the difficulties and to sustain a viable agricultural economy on such a scale. That scale is worth emphasizing: During its period of maximal development in the Byzantine era, the system of runoff farming encompassed practically all of the usable land in the northern Negev highlands.

The Negev runoff farmers apparently did more than merely gather natural runoff. We have clear evidence that they actually tried to induce more of it. The hillsides in the Negev, as in many other deserts, are naturally strewn with a pavement of stones and gravel, and this covering inhibits and detains the flow of runoff over the surface. The ancient Negevites deliberately cleared the stones off the slopes and thus smoothed the surface and exposed the finer soil to facilitate the formation of a self-sealing crust. Consequently, we find countless heaps, mounds, and strips of gravel on many hillsides, particularly in the vicinity of the old towns of Shivta, Ovdat, and Nitzana. Recent field trials in that region have shown that the practice of removing the surface gravel can increase the runoff yield by 8–20 percent.[7]

The ancient Negev dwellers also carried out larger-scale works to divert floodwater from regional streams onto adjacent flat lands. However, such works were inherently more expensive to construct and maintain. Moreover, because of the totally unpredictable and occasionally violent nature of the flash floods, the harnessing of such floods was fraught with much greater risk than the handling of small and controllable flows off hillsides.

The ancient civilization of the Negev, however remarkable, is not entirely unique. A completely different civilization, quite removed in space and time from that of the Nabateans but having to contend with similar environmental conditions, was that of the Anasazi (Pueblo) Indians of the American Southwest. The civilization of the Anasazi (meaning "ancient ones" in Navajo) developed from about 100 C.E. in the area where the boundaries of Arizona, New Mexico, Colorado, and Utah intersect. The early Anasazis, noted for their expertise as basket weavers, supplemented hunting and wild-seed gathering with the cultivation of maize, pumpkins, and beans. Later, hunting and gathering were abandoned and agriculture became the major occupation, and it included the production of cotton. The Anasazi developed a system of runoff utilization that resembled that of the Negevites to a remarkable degree.[8]

Considerable archaeological evidence found in the Chaco Canyon area of craggy northwestern New Mexico indicates that the Anasazi had a comprehensive plan for controlling surface water,[9] based on planned communal endeavor as well as practical engineering skill. The water management assembly included dams, reservoirs, canals, ditches, and water diversion walls, all of which were constructed to channel runoff water to farming terraces, garden plots surrounded by grid borders, and fields. The capture and diversion of surface runoff water was carried out for domestic as well as agricultural purposes. The Anasazi used the flat mesa tops as catchments, from which they could gather surface water from summer downpours and rapid spring thaws to supply irrigation systems in the bottomlands. The local presence of extensive outcroppings of smooth bedrock, almost devoid of moisture-absorbing soil, facilitates the collection of runoff. Practically all the rain that falls on such exposures is channeled into the canyons, thereby providing maximum amounts of water from even minimal rainstorms. Even today, it is an amazing sight in this dry desert to witness water rushing from the top of the canyon walls in a series of miniature waterfalls after a summer thunderstorm.

Commonly the Anasazi built earthen dams across the arroyos—the Spanish-derived name for a gulch or creek, equivalent to the Near Eastern term *wadi*—at the mouth of small canyons to impound the runoff in small reservoirs. From them, they built canals to direct the water to fields and garden plots situated in the wider sections of the valley bottom. They used stone headgates and ditches to distribute the water during or immediately after each runoff. The ingenious system of runoff farming, combining mesa-top water catchments with arable canyon lands, resulted in a period of population growth and cultural enrichment. To tide themselves over occasional drought seasons, the Anasazi established many large, well-constructed storage bins for grain and other durable foods.

The Anasazi civilization apparently began to decline in the twelfth century. Not much later, these remarkable people abandoned their great cliff houses and storage pits and the ceremonial chambers called *kivas,* and no one to this day is quite sure why. Several explanations have been offered. One is that the demise may have resulted from the incursion of nomadic Navajo and Apache tribes from the north. It is unlikely, however, that small bands of impoverished nomads could have subdued the well-constructed, densely populated Anasazi establishments. Another supposition is that agriculture and deforestation of the catchments led to depletion of timber resources and to accelerated erosion, arroyo cutting, and subsequent lowering of the water table, all of which might have curtailed agricultural production. Continuous irrigation and soil exploitation without any form of soil enrichment or adequate drainage may have caused the land to choke with salinity or alkalinity. Pestilence and natural catastrophes have also been postulated as possible contributors to the Anasazi's demise, but no clues to such calamities have come to light.

More plausible is the possibility that the Anasazi succumbed to the worst enemy of all people living in arid regions—a severe and prolonged drought. Paleoecological research has shown that the entire region suffered a devastating drought between the years 1130 and 1190. Such a lengthy, widespread lack of moisture would have brought dire consequences to the people so directly dependent on timely rainfall. The Anasazi system of runoff farming and grain storage was capable of coping with short-term periods of dryness, but probably could not survive a 60-year span of water deficiency. Such an extremely long drought would have been especially debilitating if it were preceded by an extended period of abundance and prosperity, during which the population might have grown beyond the number that could be sustained indefinitely in such a parched region.

Whatever the circumstances attending the decline of the Anasazi, or of their distant predecessors the Nabateans, the basic principles of their pioneering methods of land and water husbandry may well be relevant today in many of the desert fringelands around the world, where more people than ever are now struggling with the age-old problem of aridity.

Very great rivers flow underground.

LEONARDO DA VINCI

15

TAPPING THE
UNDERGROUND
WATERS

A N ASSURED SUPPLY OF WATER, primarily to satisfy the needs of humans and livestock and secondarily to provide for the irrigation of crops, is obviously the foremost condition for settlement in all arid regions. Some arid regions receive "exotic" water, originating from more humid regions, either as rivers or in the form of groundwater flowing in aquifers. The latter are porous subterranean strata that are saturated with water, and can be tapped by means of wells.

Ancient civilizations were aware that such waters exist underground. The Bible mentions "fountains and depths, springing forth in valleys and hills" (Deuteronomy 8:7). Writing in the first century A.D., Vitruvius showed a profound understanding of hydrology:

> Water is to be sought in mountains and northern regions, because in these parts it is of greater quality and is more abundant. . . . Valleys between mountains are subject to much rain, and because of the dense forests snow remains longer under the shade of the trees and the hills. Then it melts and percolates through the interstices of the earth and so reaches the lowest spurs of the mountains from which the springs flow and burst forth.

120

Tapping groundwater requires the digging of wells. The easiest place to look for groundwater is along creek beds, where the seepage of flood waters may form a water-table perched over an impervious layer at some shallow depth. The digging of such wells has always been a highly competitive enterprise. Witness, for example, the story of Isaac as related in the Book of Genesis:

> And he had possession of flocks and of herds and a great household, and the Philistines envied him. Now all the wells which his father's servants had digged in the days of Abraham his father, the Philistines stopped them, and filled them with earth. . . . And Isaac's servants digged in the creek and found there a well of living water. But the herdsmen of Gerar quarreled with Isaac's herdsmen saying: 'The water is ours!' . . . And they digged another well, and they contested over that one also. . . . And he moved from thence, and digged yet another well, and they contested not over it. . . . And he went up from thence to Beersheba . . . and pitched his tent there, and there Isaac's servants also digged a well. (Genesis 26:6, 12–23, 25)

Some of the ancient wells were quite deep and were fitted with steps on the inside by means of which a person descended and, after filling a jar, returned to the surface. This method of obtaining water is suggested in the story of Eliezer, the servant of Abraham, who went in search of a wife for the young Isaac. He met Rebekah at a well, and she "went down to the well, and filled her pitcher, and came up."[1] More poignant is the story of how the banished Hagar, finding herself without water in the desert, "cast the child [Ishmael] under one of the shrubs . . . for she said: 'Let me not look upon the death of the child. . . . Then God opened her eyes, and she saw a well of water."[2] The Moslems locate that well, which they name Zemzem, in Mecca, and consider it the holiest source of water in the world.

Of all the many ancient types of wells, none are more interesting and ingenious than the chain wells found throughout the Middle East but most typically in Iran (where they are called *karez* or *qanat*), and in North Africa (where they are known as *foggara*). In ancient Persia, where they apparently originated more than 2,500 years ago,[3] they served as the principal method of supplying water to villages and towns. Herodotus, in writing about the wars in Persia, described the way towns could be subdued by filling in their wells, thus plugging their water supply tunnels.

Chain wells are long, nearly horizontal underground tunnels designed to tap the groundwater at some higher elevation, generally at the foot of the mountains, and lead it to an outlet in the valley. The system consists of one or more mother wells drained laterally through a gently sloping tunnel that emerges at ground surface some distance from the source. In some cases, these tunnels are tens of kilometers long.

To provide access to the tunnel and a way to dispose of the excavated material, as well as ventilation for the diggers, there are regularly spaced vertical shafts dug from the surface to the tunnel. These shafts (typically 15–25 meters apart) not only facilitate the initial construction of the tunnel, but also allow the later removal of blockages that may result from occasional collapses of the unsupported sides or top of the tunnel. The surplus excavated material is deposited around the mouth of each shaft, forming a circular mound. On the surface the whole system looks like a long row or chain of hollow mounds running from the foothills down to the valley floor. The mother wells may be more than 100 meters deep. They, and the upper part of the tunnel that lies below the water table, serve as seepage inflow galleries, often with many side branches to increase the volume of water that is tapped. Extraction of water from the underground by means of *qanats* has always been a chancy enterprise. Even at best, qanat systems require maintenance to prevent clogging of the wells and tunnels.

Qanats are limited to sloping lands, usually alluvial fans of gravel outwash at the foot of a mountain range, where the seepage of floodwater forms a water table in the porous sediment resting on a relatively impervious bedrock. The groundwater drains naturally at a slow rate toward the valley floor, where it approaches the surface or emerges above ground and is generally salinized owing to progressive evaporation. At the source, however, the water is normally fresh. The purpose of the qanat system is to tap the fresh water and convey it by gravity to where it is needed for human consumption and irrigation.

When I first visited rural Iran in the late 1950s, there were *qanat* systems operating everywhere, and I had the chance to observe the *qanat* diggers at work. They were a special breed, an ancient and honorable caste proudly practicing an exclusive and hazardous craft that had been passed from father to son for countless generations. Into the shafts and tunnels and secret passageways they would crawl and burrow, molelike, with nothing more than a rope, a basket, and a hand spade. Only one man could dig at a time, and wait for the excavated material to be removed by means of bags lowered through the vertical air shafts. To illuminate the tunnel, the diggers had to rely primarily on reflected light. From time to time, the unconsolidated earth would collapse on them, and not a few were buried alive.

When I returned to Iran only 20 years later, many of the qanat systems appeared to have been abandoned, victims of modern drilling machines, power pumps, and the illusion of cheap petroleum fuel. The old chain-well system was sustainable: it could not deplete the groundwater or lower the water-table progressively. Modern wells, however, can easily overdraw water.

And what happened to those formerly indispensable and proud members of the ancient order of *qanat* diggers? Deprived of their traditional livelihood, they probably had no recourse but to gravitate toward the city, there to blend with and vanish into the amorphous crowds of a new industrial society.

Apart from Iran, chain-well systems also still operate in quite a few locations in neighboring Baluchistan (Pakistan), Afghanistan, and northern India. The *qanat* system was also introduced to North Africa and even to Spain. However, in those regions—as in Iran—this ancient method of tapping the underground waters, though tried and true, seems destined to fall prey to the convenience of the modern tubewell and motorized pump. A system that has for so long operated dependably and stably is now giving way to an alternative that will doubtless produce a flush of prosperity. But it seems fated to be followed by eventual decline due to aquifer depletion.

Examples of failures to husband water resources effectively abound throughout the history of many countries. One sad example is the fate of Fatehpur Sikri, the beautiful capital built in northern India in the late sixteenth century by the Moghul Emperor Akbar the Great. He spared no expense as he ordered the best architects and artisans to design and erect imposing walls and ornate palaces. Akbar was a great leader, strategist, administrator, and philosopher. But he was not a great hydrologist. He had not located his city next to a dependable source of either surface water or groundwater. Rather, he chose to rely on an artificial lake, designed to serve as a reservoir and to supply water for his splendid capital. The reservoir never filled to capacity, and the water that did gather in it soon evaporated or seeped into the porous ground. Less than two decades after its completion, notwithstanding the magnificence of its much-acclaimed architecture, Fatehpur Sikri was abandoned entirely, for no other reason than the simple lack of water. As we shall see, however, the folly of Fatehpur Sikri is not merely an historic curiosity, for it is being repeated today in too many places. Akbar the Great could at least plead that he had acted out of ignorance.

The knowledge of man is as the waters,
some descending from above, and some
springing from beneath.

FRANCIS BACON, *Advancement of Learning*

16

FARMING THE
WETLANDS OF
MESOAMERICA

WETLANDS TYPICALLY OCCUR at river estuaries and deltas, along low-lying coasts, in valleys, and even on high mountain plateaus. Some of the earliest civilizations of the Old World, including those of Mesopotamia and Egypt, began in wetland environments. They engaged in fishing. They used reeds to build huts and rafts, to make paper, and to weave baskets and mats. Eventually, they developed a form of agriculture specifically adapted to wetland conditions.

Nowhere, however, has the culture of wetlands been more extensive than in Central America. There, the remains of complex societies have been found, with agricultural and urban components that rank with the societies of the ancient Near East—Mesopotamia and Egypt. The remains reveal not only the ruins of cities and ceremonial centers, but also the agricultural base of those societies and its environmental context.

The Maya inhabited what are now parts of Guatemala, Mexico (Yucatan), Belize, and the fringe of Honduras, mostly in the lowlands but with some extensions into the upland areas that interrupt the plains.[1] This civilization began to emerge around 2,000 B.C.E., reached its zenith during the early part of the

125

first millennium c.e., and collapsed between 800 and 900 c.e., with a sudden decline of population and desertion of cities.

Among the chief elements of the original vegetation of the area were swamps or wetlands dominated by buttressed trees; monsoon forests on well-drained sites and on clay-filled valleys; ponded depressions; orchard-like savannas; and open grasslands. The early Mayan period was apparently dominated by slash and burn agriculture, with maize grown mainly on the lowlands. In the classical era, Mayan agriculture was more broadly based and consisted in growing sweet potato, manioc, cassava, yams, beans, squash, and cacao, in addition to maize.

An important feature of Mayan agriculture was large-scale alteration of the land surface to create an extensive series of raised fields, called *chinampas,* in areas of swamp and river floodplain. The soil was dug to form alternating canals and raised beds. Each year the act of dredging and maintaining the canals added more fertile silt to the raised beds. The canals harbored fish and turtles, and the raised beds served for growing crops.

In the upland areas which interrupt the lowlands, large tracts were terraced for erosion control. There is evidence that soil was carried upslope within the terrace systems and perhaps even from the inundated lowland areas. Altogether, the methods of lowland and upland agriculture constituted a more or less total system of land use, based on comprehensive soil and water management.

Could the problem of intensification of agriculture be the cause of the Mayan collapse? Had the population grown to such an extent that the land could no longer sustain it? Had the leaching and erosion of upland soils and the consequent silting of lowland marshes constituted the hidden scourge? Did Mayan society reach the point at which it could no longer keep up the nutrient cycle? Or was there some other factor less directly related to environmental degradation? As yet, we do not know the answers to these questions.

Among the most striking and best preserved examples of adaptation to wetlands are the remains of systems found in the extensive flood plain of the Rio Magdalena near the Caribbean

of Peru, where they cover more than 200,000 hectares. From
the air, the land for many miles resembles an expanse of corduroy
cloth, the corrugations being an alternating pattern of elongated
raised soil beds and shallow water-filled canals. Millions of these
parallel beds have been found over huge land areas throughout
central Latin America. Present-day Peruvian Indians call these
corrugations *waru-waru* and believe them to be the works of a
revered race of archaic ancestors who emerged out of the mists
of pre-Inca times.

The raised-bed farming system has recently been studied
by archeologists.[2] It appears to have been started as long ago
as 1,000 B.C.E., and to have flourished well into our millennium.
There is uncertainty over just when and why this farming method,
and the culture associated with it, ended, apparently before
the Spanish invasion in the sixteenth century. We do know,
however, that this mode of farming lasted for a very long time,
and that it provided a stable production base, relatively immune
to the hazards of flood, drought, and frost. In fact, the same
methods are viable even today. Recent experiments duplicating
the old system of soil and water management have yielded bumper
crops of potatoes (a staple crop in pre-Columbian times as well
as today) and of other crops in areas that are basically unsuited
to modern mechanized and chemicalized agriculture. So it appears
that the old system might well serve as a model for contemporary
development.

The raised-bed system found in Colombia and Peru was
similar to that of the Mayans. It consisted of a series of elongated,
rectangular platforms of piled and flattened soil, raised above
water level to a height of perhaps one meter, so as to provide
a drained and aerated rooting zone. The soil material for these
platforms was obtained by digging it from the inter-bed strips,
which thereby became shallow canals. The platforms were usually
5 to 10 meters wide, and anywhere from 10 to 300 meters
long. They were usually built perpendicularly to the main water
courses, apparently to facilitate the inflow of water to, and outflow
from, the system of inter-bed canals.

Soils in wetlands are naturally rich in organic matter and therefore tend to be fertile. The soil of the raised beds was thus fertilized at the outset, and then repeatedly refertilized during routine maintenance of the beds, by the plant residues taken with the soil dug up from the canal bottoms. Crops planted on the raised beds were watered both from the bottom, by automatic capillary movement from the canals and the high water table, and artificially from the top by withdrawing water from the canals and pouring it as needed over or between the plants. This irrigation method also helped to provide nutrients, since the water in the canals normally contained an abundance of nitrogen-rich algae, as well as other fertile sediments and remains of plants and animals.

In addition to the natural advantages of ready irrigation and fertilization, the raised-bed system also provided protection against frost by regulating the microclimate. Water in the canals would naturally tend to absorb heat by day and radiate it at night, thus moderating both hot and cold temperature extremes.

The raised-bed method is labor-intensive, but it is self-sustaining. It requires no specialized machinery or imported fertilizers. The system is relatively immune to salinity, being located in very humid regions that have long been leached of salts, and are flushed repeatedly each year by rainfall or floods.

Today these wetlands lie largely abandoned. Despite the extensive remains of an ingenious ancient agriculture that evidently provided stable and abundant support to a thriving population, these areas now appear to be a vast and forbidding wilderness of soggy marshes and floodplains, mostly beyond the pale of human habitation. But the ancient art of sustainable wetland farming might still be workable, and may yet be resurrected.

*Whoever could make two ears of corn
. . . to grow upon a spot of ground
where only one grew before, would deserve
better of mankind, . . . than the whole
race of politicians put together.*

JONATHAN SWIFT, *A Voyage to Brobdingnag*

17

THE ADVENT OF
CHEMICAL
FERTILIZERS

HUMANITY'S ABILITY TO ALTER the land has progressed through more or less distinct phases, corresponding to the development of mechanisms to control the environment: fashioning tools and weapons out of wood and stone, control of fire, domestication of plants and animals, use of draft animals, invention of implements for tillage, irrigation of river valleys, construction of water-lifting devices, terracing of hillsides, collection and utilization of runoff in arid lands, digging of wells to tap groundwater, etc. The progression has been like climbing a staircase with wide treads or plateaus, with each innovation marking a higher intensity of land utilization than the one before.

In this succession, the technological developments of the industrial revolution (starting in Europe in the late eighteenth century) outrank all those that occurred before. The two major innovations were the introduction of motorized machinery and the advent of chemical fertilizers and pesticides. Machinery and chemicals became the hallmarks of the new agriculture as it entered the modern era.

The invention of steam-driven transportation opened up the world for a global trading system in agricultural commodities.

The possibility of growing crops for export permitted a degree of specialization based on soil and climate that was unknown in earlier centuries, when international trade was limited and localities had to rely on a high degree of self-sufficiency. It thus made possible the conversion of the natural grasslands of the mid-latitudes to granaries that nourished the burgeoning urban-industrial populations of Europe and North America. The same ease of transportation of products also permitted the geographic specialization and extensive development of regions devoted to the production of wheat, rice, potatoes, tea, coffee, animal products, and various industrial crops such as cotton, sugarcane and sugarbeets, and rubber.

Human use of the soil has similarly advanced through several stages. In the first stage, an ecological balance was maintained because shifting cultivation, or low irrigation intensity, did not erode the soil or deprive it of nutrients, nor infuse it with excess water and salts. In the second stage, soil fertility declined because the soil's nutrients were mined without replacement, or the soil was damaged by erosion, compaction, or salinization. In the third stage, humans have learned to conserve the productivity of the soil by returning the nutrients removed, avoiding damage to the soil's structure, and protecting against erosion and salinization. That third stage is yet to be developed and applied fully.

The traditional farming systems generally had a period of fallow in the cropping sequence, to help restore soil fertility. The Biblical injunction, for example, required fallowing the land every seventh year, a sabbatical year called *shmitah,* to let the land rest and rejuvenate. From at least the time of Cato the Censor (234–139 B.C.E.), the Romans were also aware of the need to boost soil fertility by means of fallowing, as well as by crop rotation, liming acid soils, and adding manure. In medieval Europe, between one-third and one-half of the arable land was left fallow.

However, increases in population density gradually led to a reduction in the fractional area left fallow, until the custom of fallowing practically disappeared. Spreading of animal manures in the fields, as well as the inclusion of leguminous crops on a rotational basis, helped to add nitrogen—a principal nutrient—to the soil. Such legumes as clover, beans, and peas can improve soil fertility because of their symbiotic association with specialized bacteria that attach themselves to plant roots and that can absorb elemental nitrogen from the atmosphere, a remarkable feat that higher plants cannot perform on their own. In the rice farming areas of Asia, the occurrence of blue-green algae in the rice paddies similarly helped to supply nitrogen to the land, and the application of organic residues, including human waste, further helped to maintain soil fertility.

As agricultural production was further intensified, with multiple cropping per year and with more nutrients removed from the fields, extensive areas began to experience a progressive loss of soil fertility resulting from the depletion of the essential nutrients. Consequently, yields began to decline. Some farmers were desperate enough to glean animal and human bones from the great battlegrounds of Europe (Waterloo, Austerlitz, etc.) in order to crush and spread them on their garden and field plots. In 1840, Justus von Liebig of Germany (called the "father" of soil chemistry) proved that treatment with strong acid increased the availability of bone nutrients to plants.[1]

Necessity impelled the development of artificial fertilizers, which are chemical substances containing—in a form readily available to plants—the elements that improve the growth and productiveness of crops. The three major nutrient elements that crops need are nitrogen, phosphorus, and potassium. (Other "minor elements" are required in much smaller amounts.[2]) Fertilizers enhance the natural fertility of the soil, or replace the chemical elements taken from the soil by previous crops. The advent of artificial fertilizers represented a revolutionary change and in effect ushered in modern agriculture. Along with the development of improved crop varieties and of better methods for the control of diseases and pests, fertilizers eventually brought about a dramatic increase in crop yields.

The first artificial fertilizer was superphosphate, invented by the English agricultural chemist John Bennet Lawes. In 1842, after long experimentation with the effects of various manures on plants, he patented a process for treating phosphate rock with sulfuric acid, to make the phosphate soluble. That year he opened the first fertilizer factory, thus initiating the chemical fertilizer industry. Lawes later founded the world's first agricultural experiment station on his own estate of Rothamsted, not far from London.

Soluble forms of potassium were found in geological deposits in several countries and could be mined and used directly, but the problem remained how to supply sufficient nitrogen to satisfy crop needs. Paradoxically, nitrogen is abundant in the atmosphere, but most plants are not able to assimilate it in its elemental form. And most crops are extremely sensitive to nitrogen deficiency in the soil.

During the latter part of the nineteenth and early part of the twentieth centuries, the major sources of nitrogenous fertilizer were the saltpeter (sodium nitrate) deposits of Chile, and the guano (accumulated dung of seabirds) deposits found in Peru. The need to mine and transport these substances across the ocean, and the frequent disruption of international trade by war, made these fertilizer sources both expensive and insecure. The long-range problem of supply was solved just before World War I by the work of Fritz Haber in Germany, who found a way to synthesize ammonia virtually from air and water, by getting atmospheric elemental nitrogen to combine with hydrogen under high pressure and at moderately high temperature. This process requires energy, generally obtained from fossil fuels.

The development of artificial fertilizers was one of modern technology's greatest contributions to human welfare. In a very real sense, it fulfilled Jonathan Swift's astute observation quoted at the head of this chapter. On the other hand, it permitted the spiral of population growth (enhanced by the development of modern medicine and the improvement of hygienic standards) to continue. Moreover, the excessive use of fertilizers eventually began to contribute directly to environmental pollution and degradation. Thus we have yet another example of how an innovation designed to alleviate one problem—if misapplied or if applied excessively—may beget other problems.[3]

PART IV

THE
PROBLEMS
OF
THE
PRESENT

And the whole land thereof is brim-
stone, and salt, and a burning,
that it is not sown, nor beareth,
nor any grass groweth therein

DEUTERONOMY 29:22

18

SALINE SEEPS IN
AUSTRALIA AND
NORTH AMERICA

ACH AND EVERY ONE of the insidious man-induced scourges
that played so crucial a role in the deaths of past civilizations
has its mirror image in our contemporary world. Salinization, erosion, denudation of watersheds, degradation of arid lands,
depletion and pollution of water resources, abuse of wetlands,
and population pressure are still with us, but on an ever larger
scale. Added to the old problems are new ones undreamed of
in past centuries: pesticide and fertilizer residues; domestic and
industrial wastes including toxic chemicals, air pollution and
acid rain; global climate change; and the wholesale extinction
of species.

Among the numerous examples of how human intervention
can lead to totally unforeseen yet fateful environmental consequences, one of the most striking is the plague known as the
"saline seep phenomenon" in Australia and North America. First
noticed in isolated locations just over forty years ago, saline
seeps have already afflicted some of Australia's choicest farmland
areas, and have become the bane of countless farmers and ranchers.
Suddenly and at first quite inexplicably, the soil here and there
began to break out with what appeared to be a series of festering

135

sores oozing a briny pus. Initially, these sores came and went, as if they were a self-arresting disease, but in time they turned into permanently sterile ugly splotches in the midst of otherwise verdant fields, and they seemed to be growing from year to year.

Subsequent research revealed the saline seeps to be the delayed result of the extensive clearing of land carried out in southern Australia as long as a century ago,[1] and it illustrates how a well-intentioned act, believed to be entirely beneficent or innocuous, can result in disaster. It also shows how nature's clock can differ from our own. The timescale governing natural processes may be too slow for us to discern for a while, perhaps for quite a while, but they proceed relentlessly, nonetheless.

Geologically speaking, Australia is a very old continent, a lone island floating in the midst of a vast expanse of ocean. Over long eons of time, ocean waves have been battering against its southern and western flanks, raising a misty pall of sea-spray all along the shore. That salty spray is wafted by ocean breezes and deposited on the land either directly or by mixing with atmospheric moisture. As a result, the rains falling on a wide swath of land along the coast, and even for some distance inland, contain small but significant amounts of salt.

The slightly brackish rain then infiltrates the soil and bathes the root zone of the native trees and shrubs of eucalyptus and acacia. As the hardy deep-rooted plants extract moisture from the soil, they leave most of the salt behind, so the salt concentration in the residual moisture gradually increases. Slowly but surely, the salinized moisture continues to percolate downward, beyond the zone of root extraction. Thus, over time, the subsoil to a considerable depth becomes charged with a briny solution, even while the topsoil above remains relatively free of excess salts by the repeated leachings of fresh rains.

Significant European settlement of Australia began in the 1850s. As the new settlers extended their domain westward from their original enclave in the southeast, they developed land for farming and for grazing. The good lands with adequate rainfall lay along the southeastern and southwestern edges of

the continent, and the task of preparing these lands consisted
primarily of clearing away the trees and shrubs. This they did
on a vast scale, even importing teams of laborers from China
to speed up the eradication of millions of trees by girdling
them—killing each tree by slicing and removing a ring of bark
from the circumference of its trunk. Thus they opened up large
tracts, which they converted into wheat fields and pastures.

The foliage of a dense stand of trees and shrubs normally
intercepts an appreciable fraction of the rain (often as much as
20 percent) and the raindrops retained on the leaves then evaporate
from the forest canopy—never reaching the ground surface—as
soon as each episode of rain comes to an end. On the other
hand, the sparse foliage of a low stand of wheat or pasture
retains hardly any water at all, so practically the entire amount
of rain arrives at the soil surface. Moreover, in the natural forest
the extraction of moisture by the deep roots and its transpiration
by the evergreen leaves continues throughout the year.

The removal of that evergreen, deep-rooted, woody vegetation
and its substitution with seasonal, shallow-rooted, herbaceous
crops resulted, therefore, in a significant decrease of the total
yearly amount of evaporation and transpiration. Consequently,
a greater fraction of the annual rainfall ended up as internal
drainage and made its way downward by percolation through
the subsoil. In due course, that deep-penetrating water naturally
came to rest over an impervious stratum at some depth below
the soil, and, over the many decades, it accumulated to form a
water table. In the process, the added water also mobilized the
salt that had accumulated in the subsoil during the ages.

Gradually, the water table rose, and with it the ancient
salt. All the while, Australia's farmers continued to reap their
bumper crops of wheat, and its ranchers continued to raise their
sheep on the rich forage of clover and grass, both groups blissfully
oblivious to what was taking place below ground.

Only about forty years ago did Australians begin to notice
the new phenomenon. Since they made their original appearance,
the saline seeps, as they came to be known, have grown in
number and size until they have begun to poison sizable sections

of land, and to salinize the rivers draining the land toward the sea. Once lush and uniformly green fields and pastures have now developed ugly and widening splotches of denuded, sterile soil, encrusted with salt. Moreover, the luxuriant vegetation that once grew along the stream banks soon began to die back as the salinity of the river water increased beyond the limit of the native plants' tolerance.

Here is a problem no one had anticipated. It had seemed altogether too far-fetched that just changing the vegetation of an area could cause it to exude salt. But now that it has happened, Australians are searching for ways to alleviate the problem. What can they do? Perhaps bring back the trees? Select agriculturally useful vegetation with deeper roots and a longer active season, so as to suck more water out of the soil and thereby reduce seepage? These alternatives are being considered at present, but in the meanwhile some spotty sections of land have become so saline that it seems impractical to try to reclaim them at all.

Saline seeps occur in North America[2] as well as in Australia. They were first recognized in the mid-1940s in the non-irrigated Northern Great Plains of the United States (the Dakotas and Montana), and the western prairies of Canada (Alberta, Saskatchewan, and Manitoba)—at about the same time they were discovered in Australia. Since that time, the saline seeps have grown steadily in extent and severity until they have become a serious problem. So far, saline seeps in these regions have taken out of production nearly one million hectares, and their expansion still continues. Moreover, the high salt content, particularly of nitrate and metal ions, of the saline water is believed to be responsible for a growing number of livestock, wildlife, and fish kills.

The saline seeps of North America occur mainly in an arid to semiarid region with a precipitation range of 250–450 mm per year. These festering seeps are wet nearly all of the time; they exude a briny discharge, and often develop an encrustation of white salt. They normally form along the footslopes, but not necessarily at the lowest points in the landscape. Tractors bog down or cut deep ruts in these spots, and water often seeps into the ruts and remains there for long periods of time. Crop growth is suppressed, and in time these spots may become completely sterile.

The phenomenon in North America is not entirely analogous to the one in Australia: in America, the original vegetation was grass, not trees, and the source of the salts was not sea spray but underlying deposits of salt-laden marine shales, originating in a much earlier geological era. However, the process by which the seeps became manifest was similar for both continents, in that it resulted from human interference with a preexisting ecological equilibrium. Under the native sod of the North American prairie, only 1–4 percent of the annual precipitation normally reached the groundwater flow system, but this fraction was increased to about 7–15 percent under the crop-fallow farming system which has been practiced in the region for many decades. The purpose of the summer fallowing is to store moisture in the soil for subsequent crop growth, but during the wet season the moisture intake often exceeds the soil's storage capacity and the excess moisture drains downward. With a mean annual precipitation of about 350 mm, the amount of groundwater recharge, which was less than 15 mm under the native sod, more than doubled to 25–55 mm after the land was put under cultivation. This augmentation of the groundwater, resulting in a progressive rise of the water table, initiated and accelerated the formation of the seeps. And, as the "new" water mixed with the "old" groundwater, it brought up the salts the latter contained.

The glaciated landscape of the Northern Great Plains has a swell-and-swale topography, with numerous depressions that cause water to pond for extended periods. Post-glacial erosion has accentuated the local elevations and inequalities. Owing to this topographic relief and the hydraulic gradients underlying it, the raised groundwater tends to move laterally until it breaks out at the surface, generally along the lower reaches of sloping ground. The naturally occurring salts in the groundwater are augmented by fertilizer residues and by the soluble products of the decomposition of soil organic matter leached from cropped fields. As a result, the upper layers of groundwater underlying cultivated land contain high concentrations of various salts, in cluding nitrates, as well as toxic elements such as selenium. The poor quality of the groundwater has made it undesirable for consumption either by livestock or humans. Data indicate that enough salt is stored in the top 6 meters of the aquifer to maintain the existing saline seeps for another 25 to 100 years.

What can be done to control or alleviate the saline seep problem in North America? Here, as in Australia, two approaches are possible: agronomic and engineering. The agronomic approach aims to use all or nearly all the precipitation by eliminating the practice of fallowing and by growing deep-rooted and long-season crops such as alfalfa, able to utilize essentially all the precipitated moisture. By thus reducing water seepage below the root zone, the amount of water moving to the saline seeps might be reduced. If the water table can thus be lowered, many seeps will presumably disappear. The engineering approach on the other hand, attempts to intercept the saline groundwater artificially before it reaches the surface, and thus to maintain the water table at a safe depth. A drainage system (consisting of "mole" channels formed in the subsoil by a special machine, or of installed perforated tubes called tile drains) is put in place where possible, to catch and convey the water to a safe disposal site. Finding a safe outlet is itself a problem, as the high salt content of the effluent poses a danger of contamination to streams and ponds.

On the positive side is the proven fact that an appropriate combination of agronomic and engineering measures, though expensive, can alleviate the saline seep problem in many places, and can serve to reclaim the affected soils within perhaps 10 years. In practical terms, however, the question is whether the affected farmers will have sufficient economic incentive to actually implement reclamation of the saline seeps. Otherwise, the control of saline seeps might be like the control of erosion, for which the technology is available but for various reasons is seldom applied.

. . . as water spilt on the ground,
which cannot be gathered up again . . .

II SAMUEL 14:14

19

THE PROMISE AND
PERIL OF IRRIGATION

IRRIGATION IS THE PRACTICE of supplying water artificially to permit farming in regions that are arid, and to offset drought in those that are semiarid or semihumid. Irrigation can thus promote production in areas where it would otherwise be impossible or impractical. Even where, on average, the total annual rainfall might be sufficient for cropping, it may be unevenly distributed during the year and vary from year to year, so that traditional rainfed farming is a high-risk enterprise and only irrigation can ensure a stable system of production.[1]

Irrigation can prolong the effective growing period in areas with dry seasons, thus allowing multiple cropping—two, three, and sometimes four crops per year. With the security it provides, additional inputs that further promote high yields (fertilizers, improved varieties of seed, pest control, and better tillage) become economically feasible. Irrigation reduces the risk of these expensive inputs being wasted by drought. Altogether, the potential productivity of irrigated land can exceed that of nonirrigated land several fold, thanks both to increased yields per crop and to multiple crops each year.

The promise of irrigation is great: it can literally make the desert bloom. But so are its potential problems. So pervasive and inherent are these, in fact, that some critics doubt whether irrigation can be sustained in any one area for very long—and they have much evidence to support their skepticism.

From its early and primitive antecedents in the river valleys of the Near East some seven millennia ago, the practice of irrigation has gradually helped increase the farmer's control over crop, soil, and even weather variables. Although only partial control is possible even today, modern irrigation can be a highly sophisticated operation, involving the simultaneous manipulation of numerous factors of production. And yet, problems remain, and in many cases they are more serious than ever. As in ancient times, short-term gains in production resulting from the introduction of irrigation often lead to long-term loss in the form of water-resource depletion and pollution, as well as of soil degradation. These problems are universal and are no less acute in highly developed countries than in less-developed ones.

The total area of land being irrigated in the world was about 149 million hectares in 1965.[2] By 1975, irrigated land had expanded to include about 223 Mha. Now, in 1990, the land area under irrigation exceeds 270 Mha. Five countries—China, India, Pakistan, the Soviet Union, and the United States—account for about two-thirds of this area. Numerous other countries—notably Egypt, Sudan, Iraq, Jordan, and Israel, to mention just a few—are vitally dependent on irrigation. It is estimated that irrigated land, presently comprising little over 15 percent of the world's total cultivated area, produces well over 30 percent of the global agricultural output. With population growth and the pressure to improve living standards and provide food security everywhere, there is a tendency to expand irrigation wherever possible. However, the future of irrigation is threatened by the constraint of dwindling water supplies, and by the twin menaces of waterlogging and salinization.

It is a disconcerting fact that irrigated farming in very many areas falls far short of achieving its potential. Even more disconcerting is the fact that extensive irrigated areas have undergone deterioration to the point where they have either already been

abandoned or seem destined to be abandoned. Is the problem
intrinsic to the principle of irrigation as such, or merely to the
careless practice of it? Must irrigation necessarily become self-
destructive sooner or later, or can it be sustained in the long
run?

Experience leads me to believe that the problem lies in mismanagement. What is at fault is the unmeasured and generally
excessive application of water to the land, with little regard
either for the real cost of the water—in contrast with its arbitrarily
set price, which frequently is too low—or for the potentially
destructive processes thereby set in motion. Another frequent
and closely related fault is the failure to provide for drainage,
and to manage the salts as well as the water so as to prevent
the insidious process of soil salinization.

It is a universal fallacy to assume that if a little of something
is good, then more must be better. In irrigation, as indeed in
many other activities, just enough is best, and by that we mean
a controlled quantity of water just sufficient to meet the requirements of the crop and to prevent the accumulation of salts—
no less and certainly no more. Applying insufficient water is
an obvious waste, as it fails to produce the desired benefit; on
the other hand, applying an excess can be still more harmful,
as it tends to impede aeration, leaches nutrients, induces greater
evaporation, raises the water table, induces salinization, and
greatly increases the cost of drainage.

Since the installation of a complete groundwater drainage
system is a complex, exacting, and above all expensive operation,[3]
it is altogether too tempting to start new irrigation projects
while delaying the installation of drainage as long as possible.
Since, however, the gradual rise of the water table is not readily
evident at the surface, that attitude often means that planners
wait until the degradation process is so severe that the project
may be too difficult or expensive to rehabilitate.

Even where groundwater drainage of irrigated land is feasible,
there remains the problem of its disposal. In some cases, the
drainage water can be reused for irrigation. In other cases, the
drainage effluent is too brackish to be so used. Intensive irrigated

agriculture releases not only salts but other chemicals as well, including residues of fertilizers and pesticides. All too often, the control of such pollution consists of exporting the pollutants— and the problem—to other sites, where the damage inflicted may be considerable and must be borne by someone other than its originators. The primary requirement of any drainage system, therefore, is that there be an outlet for the proper disposal of the effluent. This requirement may involve regional environmental considerations reaching far beyond the confines of any particular irrigation district.

The common practice of dumping the drainage effluent back into the river merely serves to salinize the water supply—by now diminished, in any case—upon which depend the less fortunate water users who are located downstream from the point of discharge. If they, too, drain their fields in similar fashion, the river will become progressively saltier and its lower reaches may become unfit as a water source for either humans or crops. The river then turns into a saline stream, with consequent effects upon its associated aquifer and estuary, or upon the lake or bay into which it discharges. If, in addition to agricultural drainage, municipal and industrial effluents are also released into the same river, it becomes in effect an open sewer.

Irrigation systems may have other negative, albeit unintended, effects on the societies and countries they are supposed to benefit. In the rural areas of many of the poorer countries, the irrigation system may constitute the only water supply to a village or a district. Unlined canals often serve as the source of water not only for agriculture but for domestic use as well— for drinking, washing, bathing, recreation, and the disposal of waste. Moreover, open field canals also attract animals such as water-buffaloes that wade and wallow in the canals and disrupt their banks. Under such circumstances, the water becomes very polluted and spreads a variety of diseases. Riparian vegetation growing along canal banks may shelter snails that act as vectors for the spread of schistosomiasis (bilharzia), a debilitating disease that has become endemic among riverine populations in parts of Africa and Asia. Other diseases that have been associated with the spread of irrigation include malaria, river blindness, cholera, and various diarrheas. The number of people affected by these maladies runs into the millions.

As major water users, some large-scale irrigation projects operate in an inherently inefficient way. Where water is delivered to the consumer only at fixed times, and charges are imposed per delivery regardless of the actual amount used, customers tend to take as much water as they can while they can. This often results in over-irrigation, which not only wastes water but also causes project-wide problems connected with the disposal of excess water, waterlogging of soil, leaching of nutrients, and elevation of the water table. In the absence of reliable statistics, it has been estimated that the average application efficiency— the fraction of the water delivered that is actually used by the crop—is well below 50 percent, and in many places may even lie below 30 percent. Since it is proven that efficiencies as high as 90 percent can be achieved, there is obviously much room for improvement.

Particularly difficult to change are management practices that lead to deliberate waste, not necessarily because of insurmountable technical problems or lack of knowledge, but simply because it appears more convenient, or even more economical in the short run, to waste rather than to conserve water. A typical situation is when the price of irrigation water is lower than the cost of the labor or equipment needed to avoid over-irrigation. Very often the price of water does not reflect the true cost of providing it but is kept deliberately low, perhaps for political reasons, by government subsidy. The cost of water may be distorted even in the absence of governmental subsidy. For example, when users draw water from an aquifer in excess of the rate of annual recharge, the cost of pumping may be only a small fraction of the cost of replenishing the aquifer after it has been depleted. But by the time the aquifer is emptied, the over-drawers have made their profit and gone, leaving the problems to future generations.

Where open and unlined distribution ditches are used, major losses of water can occur through uncontrolled seepage and evaporation, as well as transpiration by riparian vegetation. Even pipeline distribution systems do not always prevent loss, and may leak water from loose or corroded joints or ill-maintained valves. Sometimes such losses of water are not immediately apparent, and they may escape economic valuation entirely.

All these problems with irrigation are not merely theoretical: they occur in actuality, and on a very large scale.

A case in point is the Indus River Basin in Pakistan and parts of India.[4] Before systematic region-wide irrigation was instituted near the turn of the century, the groundwater table in this great basin, which stretches from the foothills of the Himalayan Mountains to the Arabian Sea, was generally scores of meters below the soil surface, and the aquifer was in a state of hydrological equilibrium.

When large-scale irrigation was introduced, an extensive water distribution network was established to form what is probably the largest continuous irrigation district in the world, covering about 15 million hectares. The irrigation system integrates the Indus River and its major tributaries, 3 principal storage reservoirs, 19 barrages and headworks, 43 canal commands, and about 90,000 watercourses. The total length of the canals is about 56,000 kilometers, and there are more than 1,600,000 km of watercourses and farm channels. The system was planned to spread water over as large an area as possible. The efficiency of the distribution system, however, is very low: studies show that 35–45 percent of the water diverted from the rivers by the canals never reaches the farmers' fields. Moreover, the field irrigation efficiencies are even lower—possibly about 30 percent. Since the natural drainage is inadequate and artificial drainage was not provided in the early decades of the project, percolation from the canals coupled with extensive over-irrigation led to a rapid rise in the water table—about 30 to 40 centimeters per year.

Areas along the main canals were the first to become waterlogged, and waterlogging subsequently spread to contiguous areas. By 1960, the water table was within 3 meters of the soil surface (a threshold depth at which the twin menaces of waterlogging and salinization can become acute) under about half the canal command area, and within one meter of the surface over a tenth of the area, thus causing problems of waterlogging and salinization in about 1 million hectares. By 1980, the situation had grown much worse. The groundwater was within less than 3 meters of the surface underneath 55 percent of the total irrigated area. Salinization now affects an estimated 5 million hectares. One need only travel overland in the Punjab ("the Land of Five Rivers") to see that the extensive carpet of green is now interrupted not just by spots of salinity but indeed by vast and ugly stretches of waterlogged land that has become thoroughly unproductive.

This alarming trend led the government of Pakistan, with international assistance, to undertake an expensive large-scale drainage program without which, undoubtedly, some 50 percent of the irrigated land would by now be unfit for normal cropping. This program includes the construction of a regional drainage canal, and also encourages private users to install tubewells to pump up water from the aquifer in places where the groundwater is of good quality. Thanks to this program, the water table has ceased rising in large areas and is even being lowered in some. In other areas, however, the rise in the water table still continues, though more slowly. Hence the drainage work must be expanded and improved, or else some 50 percent of the remaining irrigated land will turn into a series of brackish marshes. What makes the drainage canal particularly expensive is that most of Pakistan's irrigated land lies at a great distance from the sea, and the downward gradient of the land toward the sea is extremely small (less than 1:5,000). The drainage problem in Pakistan's Punjab is shared by irrigators in India's East Punjab, but lack of cooperation between the two countries greatly complicates its alleviation.

Egypt depends entirely on irrigation to feed its 55 million people. The country is basically a desert, with practically no agriculturally effective or reliable rainfall. Its sole source of water is the Nile, whose flow varies from year to year. It was to regulate that flow, and to protect the country from the effects of the periodic droughts that afflict eastern Africa, that the Egyptians in the 1960s undertook to build the Aswan High Dam, with the help of the Soviet Union. That dam has had a profound effect on Egypt's water regime. In some respects, it has been a boon to the country's development, but in other respects it has been a bane. The larger environmental issues and controversies connected with this dam (such as the blocking of the river's silt which had previously helped to restore the fertility of Egypt's irrigated soils) are beyond our scope. Suffice it to say that the water stored by the dam in Lake Nasser has indeed enabled Egypt to regulate its water supply, and to escape the potentially dire effects of the severe droughts of the 1970s and 1980s. The year-round availability of water has allowed increased irrigation, with its obvious benefits and not-so-obvious drawbacks.

Since the water level in the Nile is now prevented from pulsating up and down annually, as it did throughout Egypt's long prior history, the river no longer serves as effectively to lower the water table and flush out the salts. The water table has even tended to rise in the lands along the river, and particularly in the Nile Delta, and Egyptian farming has begun to experience the plagues of waterlogging and salinity to which it had seemed so long immune. The diversion of water to the "New Lands Program,"[5] designed to expand the area under irrigation by reclamation of desert tracts on the fringes of the flood plain, also has had the effect of exacerbating the drainage and salinization problem in the "old lands," which are at a lower topographic elevation and hence receive lateral groundwater flow from the new lands. Egypt must now invest in developing an expensive artificial drainage system, lest it lose more and more of its best land to soil degradation.[6]

Intensified irrigation and the maintenance of the water level in the Nile have afflicted Egypt with still another salinity problem beyond its irrigated lands. Instead of being flushed out, as they were in the past, by receding flood waters, the salts that now remain in the groundwater are infused by capillary action into the porous soil and rocks. Water piped in to supply the needs of the expanding population of towns and villages along the

Nile, and cesspools placed underground to dispose of their wastes, have further raised the water table. As a result, there is now a constant upward seepage of salt-bearing moisture into Egypt's ancient temples and monuments, and these salts impregnate the porous stone walls.

As they say in Egypt, "salt is like a sleeping devil—only when it gets moisture does it start to act." When the moisture evaporates at the exposed surfaces of these structures, the salts recrystallize, forcing apart the grains of stone. The result is a flaking and crumbling of the ornately carved reliefs and inscriptions of Egypt's magnificent monuments. Salt bubbling up under the ancient wall paintings pushes the plaster off the walls, so that the exquisitely drawn and brightly colored portraits of the ancient kings, queens, and gods, as well as the vivid depictions of landscapes and scenes of daily life (notably including farming activities), are now deteriorating rapidly. If this deterioration continues for a few more decades, many and perhaps most of the reliefs and paintings adorning the ancient temples and graves will be erased from within, and only the blank, pulverized surfaces of walls and columns will be left.

Some engineers have proposed installing pumps to lower the underground water level, and digging trenches, filled with gravel, around the bases of temple walls to prevent water from seeping into the stone. But such a scheme is certain to be expensive and may not work as intended. Attempts have been made to remove salt stains from art on walls by covering them with wet paper and absorptive clay, but the salt deep in the walls could not be removed this way, and one month after the surface was cleaned the stains reappeared. Faced with this accelerating decay, the protectors of Egypt's priceless archaeological heritage are just as stymied by this insidious and formidable enemy— salt—as are the neighboring farmers.

The dilemma of land degradation under irrigation is not the exclusive province of the so-called less-developed countries. Problems of equal or even greater severity occur in such highly developed countries as Australia, the United States of America, and the Soviet Union.

In Australia, a serious problem now faces irrigated agriculture in the Murray-Darling river basin,[7] which covers about one-seventh of the Australian continent. It is that country's most bountiful food basket: it supports one-quarter of the nation's cattle and dairy herds, half of its sheep and cropland, and nearly three-quarters of its irrigated land. Indeed, it has been a thriving irrigation district for scores of years. But the district is suffering increasing problems of salinity and land degradation, and the effects are felt acutely beyond the agricultural district itself. South Australia, into which the river system discharges, receives the accumulated impurities that up-stream activities have added. The main impurity is salt: more than 1.3 million tons of it are delivered by the river into South Australia each year. In a dry year the concentration of salt can exceed the maximum level the World Health Organization considers to be appropriate for drinking water.

Much of the river's salt comes from the inflow of saline groundwater, caused by rising water tables. These, in turn, have been brought about by two factors—the clearing of natural vegetation, and irrigation. The original wooded vegetation, with roots extending to a depth as great as 17 meters, is extremely efficient at extracting soil moisture. Under natural conditions, only about 0.1 millimeters of the average 250 mm of rainfall escape past the root zone to recharge the aquifer 30–40 meters below the surface. However, over much of the Murray Basin, the original vegetation has long been stripped away and the recharge rates have increased significantly. Add to that the extra seepage from irrigated lands, and it is not surprising that the water table has been rising dramatically, in some areas at rates exceeding 20 centimeters per year. The water table has already reached the surface in some spots, and waterlogging is spreading. The amount of salt stored under the surface is astonishing: some 3,000 tons per hectare, or even more. If the rise in the water-table is not checked, more land will be retired from production and become salt-encrusted, lifeless terrain, and more salt will be released to the rivers until they turn into saline drains.

Soil salinization is already widespread. About 140,000 hectares of irrigated land in Victoria alone are now affected by salinity, and 400,000 more are considered salt-prone and at risk of being salinized. The government of Victoria now recognizes salinization as the single greatest threat to the state's environment. Altogether, more than 520,000 hectares in the Murray basin overlie a vast pool of salt water that has risen to within 2 meters of the surface. In New South Wales, nearly 20 percent of the irrigated land is suffering salinization, and the area affected by high water tables is expected to increase from 200,000 to about 300,000 hectares in the next decade.

How can one remedy the region's ecological ills, and achieve efficient and sustainable production while avoiding, or at least minimizing, off-site effects? One can treat symptoms or eliminate causes. Engineering schemes can stop much salt from reaching the river, by constructing channels or subsurface drains to intercept groundwater before it reaches the surface and pumping it away to evaporation ponds. But this approach cannot be sustained indefinitely—it only buys time, perhaps 30 years, before the evaporation basin fills up or the rising tide of salty groundwater overwhelms pumping capacity.

The permanent and much more comprehensive solution is to eliminate the causes, by reforesting aquifer-recharge areas and phasing out flood irrigation of land prone to waterlogging. That means ending existing pastoral and agricultural activities in large areas. Better irrigation systems are needed to reduce excessive percolation and to achieve more efficient utilization of water.

Lest we think that irrigated agriculture in the United States is faring better than in Australia, a few examples will show that this country is suffering from the very same self-inflicted problems.

The Colorado River's salinity has been rising during the last few decades as it flows downstream through the States of Nevada, Arizona, and California, in each of which irrigators draw water from it and discharge their drainage into it. The current level of salinity at the Imperial Dam is over 900 parts per million (ppm) and it is predicted to continue to rise and to exceed 1,200 ppm after the year 2000. Some years ago, the U.S. government signed an agreement that commits this country to lower the salinity of the river, which flows into Mexico. Desalination of the water is certain to be an extremely expensive operation. The alternative is to institute much more efficient irrigation systems in the area involved, and to drastically reduce the discharge of salt-laden drainage into the Colorado River. A similar problem besets the Rio Grande.

Irrigated farming in the western section of the Central Valley of California is currently facing a crisis. To avoid salinization, the region requires drainage. But where should one discharge the effluent? Since a mountain range comes between the Central Valley and the Pacific Ocean, the region's farmers originally planned to convey their drainage via a canal to San Francisco Bay. However, they encountered the strong opposition of Bay area residents, who resented the idea of their closed bay becoming the sump for agricultural drainage.

An alternative plan was then conceived: discharge the effluent into a basin called the Kesterson Reservoir. This plan was carried out, and for a while it seemed to be a great success: an artificial wetland was created that harbored a great diversity of wild flora and fauna, including numerous aquatic birds. The agricultural industry of California pointed to this reservoir with great pride as a major contribution to the state's environment. Then, just a few years later, biologists noticed a baffling and disturbing fact: many of the young birds were deformed and died before reaching maturity. Other animals were similarly affected. The culprit was found to be selenium, an element that—ironically—is an essential nutrient in small amounts, but becomes toxic at higher concentrations. It was present in trace quantities in the region's groundwater, and became progressively concentrated as the drainage water evaporated in the open reservoir. Reluctantly, region's irrigators had to abandon the Kesterson scheme. Now the large irrigation district involved is left without an outlet for its drainage, and may be doomed to suffer the fate of so many other irrigated regions throughout history unless some other solution can be found.

Irrigation in California, incidentally, is responsible for the continued existence of another body of water, a large lake called the Salton Sea, located in the Imperial Valley (once called the Valley of the Dead) near the Mexican border. Although it had been inundated repeatedly in past ages, the Salton basin had been a dry depression in recent history. In 1904, however, the as-yet-untamed, undammed Colorado River breached the banks of the artificial channel built to contain it, and flowed into that depression to form a new fresh-water lake. By 1907, the breach was repaired, and the new lake would normally have evaporated away. But, contrary to expectations, the lake has remained, and even grown, since then. What sustains it artificially is the collective drainage from the Imperial Valley, now one of the most intensively irrigated districts in America. It is also used for recreation and is a permanent feature of the valley. However, the salinity of this landlocked lake, lacking any cleansing through-flow (its only outlet is the evaporative sink of the dry desert air) has increased steadily and is now close to the salinity of ocean water.

Still another example of irrigation management is the case of the Ogallala aquifer in the Great Plains region. In the southern part of this region (including parts of Kansas, Colorado, Oklahoma, New Mexico, and especially the Texas Panhandle), water is being drawn from wells at a rate far in excess of the aquifer's natural recharge (in fact, at a rate equivalent to the entire flow of the mighty Colorado River), and the water table has been falling progressively. For some decades the farmers of this region, instead of cooperating to regulate pumping regionally and to replenish the aquifer insofar as possible, have been competing in drawing water for their own immediate profit without regard to the region's future. Since water is a fluid and recognizes no property boundaries, any individual who attempted to conserve water was in danger of having his water pumped out from under him by his less scrupulous neighbors. And so, in a case of private enterprise carried to an aberrant extreme, the entire thriving irrigation district had been engaged compulsively in the exercise of putting itself out of business. It is now forecast that within the next two or three decades the water level in the aquifer will have fallen so far as to make the cost of further pumping entirely prohibitive for irrigation.

We close by describing a major ecological catastrophe, ironically resulting from what had for some decades been held up as a shining success of irrigation development.[8] In the arid plains of Soviet Central Asia, in the vicinity of the Aral Sea, a large region was placed under irrigation without any attention to environmental constraints. But the hubris of central planners, determined to force an entire region to produce a single commodity deemed essential to the Soviet economy, has brought on its nemesis in the form of a ghastly plague that could and should have been foreseen.

With its warm, sunny climate, the central Asia region is to the USSR what California is to the United States—a rich fruit and vegetable basket. It covers only about 7 percent of USSR territory but yields more than 33 percent of the country's fruit, 25 percent of its vegetables, and 40 percent of its rice. Above all, the region produces 95 percent of the Soviet Union's world-leading harvest of cotton.

Called "white gold" by the Russians, cotton is a particularly important crop in a country where synthetic fiber production is underdeveloped, and it earns valuable hard currency in export markets (though lately China has emerged as a powerful competitor in international cotton exports). The Soviet central state planners calculated that, by massive irrigation, they could simultaneously develop one of their most economically backward regions, provide the local people with employment, and serve the European and Siberian mills with vital raw material for their textile industry.

A fragile terrain of oasis and desert has been appropriated largely to satisfy the central planners' demand for raw cotton. The sheer scale of the investment and achievement can be seen by anyone flying over the region: long lines of collective farms stretching for hundreds of kilometers into the reclaimed desert. But the central planners were ignorant of the old nomadic proverb: "Water brings life, but it can also bring death." Their plan set quantitative targets, not qualitative ones.

Although ideal for heat-loving crops, the Central Asia region is arid, so irrigation is essential. To provide the water needed for irrigation, engineers diverted the flow of the region's two rivers, the Amu Dar'ya and the Syr Dar'ya (the Oxus and Jaxartes, as they were known in classical times), both of which flowed naturally into the Aral Sea. So much of the water was siphoned off that the first of these rivers now fails to reach the Aral Sea in relatively dry years, and the second has not reached it at all for some 20 years. Deprived of replenishment, the Aral Sea, once the world's fourth largest lake, began to shrink and is now in danger of disappearing. Its level has fallen by some 15 meters and its area has shrunk by about half in the last 30 years. The once-thriving fishing industry has been devastated, and fishing villages once located on the shore are now stranded 30 to 80 kilometers inland. And, of course, the salinity of the water has risen dramatically.

Behind the receding waterline lie mudflats with millions upon millions of tons of fluffy salt and chemical residues which, picked up by the swirling continental winds, float in deadly dust clouds to destroy crops and poison the land for hundreds of kilometers around. A witness has described the scene in Biblical terms:

> The sky is covered by a salty curtain, the sun becomes crimson and disappears in the salt dust. In the entire province, not one tree grows on the land. The livestock are perishing, and the people are getting sick and dying.

Even the climate of the region seems to have changed perceptibly. If the same practices are continued for another 20 years, the Aral Sea could end up as only a small briny swamp in the midst of a sterile and uninhabitable desert of salt.

Compounding the salinity are the residues of agricultural chemicals, applied in huge overdoses in an effort to coax the greatest yields from the land in the shortest possible time. An average of 30 kilograms of fertilizer, containing one or two kilograms of toxic chemicals, is applied to each hectare of agricultural land each year in the USSR. Up to 20 times that much is dumped on Central Asia's cotton crop. In addition, toxic pesticides and defoliants have been poured on the vast cotton fields. The chemicals have seeped into the groundwater and the surface streams, poisoning the only water supply available for the region's population.

According to a report published in June 1989 in *Socialist Industry,* an official Communist Party organ, two-thirds of the people in the Karakalpak region bordering the Aral Sea suffer from hepatitis, typhoid, gastrointestinal diseases, or throat cancer. Especially affected are the children, who suffer from a variety of illnesses including anemia, rickets, and liver complaints. Infant mortality in Soviet Central Asia is as much as four times the USSR average, ranging from 46 per thousand in Uzbekistan to 58 per thousand in Turkmenistan. In some areas, more than 1 in 10 of the babies die in their first year, and many are born with deformities. Nevertheless, the indigenous population of these regions has been increasing by 3 percent per year, a rate that exacerbates the problem as it makes much more difficult the provision of alternative sources of livelihood.

The environmental crisis in Soviet Central Asia has been recognized officially, in the usual way, by the creation of a special commission. "In theory, the question is settled," said the president of the Turkmenistan Academy of Science, "but we must put it into practice." Long-proposed schemes to divert the waters of such Siberian rivers as the Ob, Irtysh, and Yenisey, and to channel them south to the Aral Sea and the desert, were cancelled by the Soviet government after years of controversy about their costs and environmental consequences. Another such visionary scheme was to break up and melt the glaciers of the Pamir and Tian Shan mountains with nuclear explosions.

A more practical approach, long overdue, is to modernize the inefficient irrigation system by lining canals, laying pipes in place of open channels, applying water by means of sprinklers or drippers instead of by surface flooding, introducing volume-controlled sluices and valves, and altogether promoting greater efficiency in the distribution and use of irrigation water. This might also involve water pricing, as well as methods of using groundwater judiciously in conjunction with surface waters. The trouble is that soil salinization has been neglected for so long that enormous quantities of fresh water will be needed just to flush out the land that has already been made saline.

In October 1988, the Communist Party Central Committee decreed that after 1991 no new irrigation projects will be allowed to draw water from the rivers feeding the Aral Sea. Moreover, by the year 2000 farmers will be required to upgrade irrigation systems to cut water-withdrawal by 25 percent. Water fees will be instituted within two years in order to induce farmers to improve irrigation and water-use efficiency, with the goal of increasing minimum flow into the Aral Sea tenfold by the year 2005. If achieved, this goal will probably preserve the sea, albeit at a diminished size.[9]

The terrible specter of the Aral disaster is an object lesson for all irrigation planners[10] and managers in arid regions. Every irrigation project should aim at increasing the efficiency of water use rather than merely increasing the supply of water to accommodate inefficient and environmentally damaging use. To increase the supply is generally more costly, and in any case it only postpones the problem. In many countries, nearly all the available water resources are already in use, so increasing efficiency is the only solution. Irrigation, as it is practiced in most countries, is terribly inefficient. Only about a third of the water diverted for irrigation is used effectively by agricultural crops; the rest is lost in direct evaporation or in excess seepage, which in turn exacerbates the problems of a rise in the water table, waterlogging, and salinization.

The greatest opportunities for water conservation come from new irrigation techniques, by which water is delivered in closed conduits and applied directly to the root zone of crops, in small volume, at high frequency, and at a controlled rate in exact response to their optimal needs.[11] It is often said that such modern methods are too costly for adoption in less developed countries, so these countries have no choice but to remain with their traditional methods of surface irrigation, conveying water in open channels and running it over the surface of the ground.

This opinion is grossly fallacious. It considers only the direct installation and operating costs, but ignores the hidden costs of water wastage and of land degradation, as well as the increased eventual costs of drainage and land reclamation. Although the latter costs are not immediate, they will occur eventually; and when they are taken into account the relative costs of modern versus traditional irrigation methods will change radically. Furthermore, there is great unrealized potential for simplifying and reducing the costs of modern irrigation methods, and of adapting them to the needs and circumstances of the less developed countries, where the relative costs of capital and labor differ greatly from those that prevail in industrialized countries.

*Thou shalt inherit the Holy Earth as
a faithful steward. . . . Thou shalt
safeguard thy fields from erosion.*

WALTER CLAY LOWDERMILK,
"An Eleventh Commandment"

20

ACCELERATED EROSION

TWO EARLY LEADERS of the U.S. Soil Conservation Service, Hugh H. Bennett and Walter Clay Lowdermilk, who surveyed the global dimensions of the soil erosion problem, wrote in the *1938 Yearbook of Agriculture:*[1] "Soil erosion is as old as farming. It began when the first heavy rain struck the first furrow turned by a crude implement of tillage in the hands of prehistoric man. It has been going on ever since, wherever man's culture of the earth has bared the soil to rain and wind." Asked to testify on soil erosion before a committee of Congress, the two of them, without a word, placed a thick towel on the committee's polished table and poured a large cup of water onto it. The towel soaked up the water. Next they removed the wet towel, and, still saying nothing to the puzzled members of the committee, poured a second glass of water on the bare table. The water splashed and trickled off the table and onto the laps of the distinguished committeemen. Every one of them then understood the dire consequences of removing soil from hillslopes.

Soil erosion is a natural phenomenon, a process by which all protruding surfaces of land are worn down gradually by the elements—only to be uplifted and re-formed time and again by geological forces.[2] The problem begins when human activity intervenes to hasten that wearing down process much beyond the rate at which nature can redress it.

A commonly cited estimate of the mean rate of soil formation in nature is that a foot (about 30 centimeters) of soil develops in about 10,000 years. This is, at best, a very gross estimate, as the actual rate of soil formation is known to vary widely depending on climate, parent material, topography, and age of the soil. In a mature soil, the natural rate of soil formation is balanced by the natural rate of erosion, called "geological erosion." In good farming practice, tillage and accumulation of organic matter may help to build topsoil at a faster rate, enough to compensate for the probable increase in erosion that typically attends the practice of crop production. Much more commonly, however, injudicious farming practice causes accelerated erosion.

The problem of accelerated erosion leading to rapid loss of topsoil, the most fertile layer of the soil, is worldwide, and growing worse.[3] It is exacerbated by modern patterns of land use and mechanized methods of cultivation. In the main, these methods were developed in Western Europe, where they resulted in great changes in the landscape, notably removal of most of the forest cover and drainage of wetlands, but in general they did not cause a progressive degradation of the land and water resources there. The benign temperate-humid climate of Western Europe, with its relatively gentle rainfall regime, spared Europe— with the notable exception of some Mediterranean sections— the extreme forms of soil erosion that occurred in other regions. But when Europeans introduced their clean-till farming methods to other regions with less benign climates, severe erosion ensued.

In the United States, evidence of erosion can be found in almost all agricultural regions. A particular example is southern Piedmont, where the soils are mostly acidic, nutrient-poor, and highly erodible, and where the rainfall tends to be highly erosive. Here, the introduction of intensive methods of cotton growing subjected the erodible soil to the often violent downpours of a region especially prone to violent tornadoes and hurricanes. The accelerated erosion was particularly severe during the middle of the nineteenth century, when plantation agriculture was practiced widely.

Examining these soils today, we can definitely discern the evidence of that erosion, although it is now difficult to find virgin sites with which the eroded soils can be compared to determine just how much of the original topsoil has been lost. The best data available suggest that the original topsoil (though it undoubtedly varied greatly in this region of uneven topography) may have had a typical depth of about 40 centimeters. A yield reduction of about 40 percent can occur when just a depth of 15 cm of the topsoil is removed. And yet sizable areas in the region have apparently lost their entire cover of topsoil, so that the B-horizon is now the layer being plowed. Whereas the A-horizon topsoil was naturally friable and had a reserve of fertility thanks to its organic matter content, the B-horizon is typically sticky and plastic, and thus inherently less favorable for planting crops.

In recent decades, crop yields have generally been increasing despite the obvious process of soil erosion. Here we must make a distinction between actual and potential productivity. The former can continue to rise even while the latter falls, simply because the detrimental effects of erosion on soil fertility can be masked for a time by increased applications of fertilizers, use of better varieties, denser plantings, more intensive pest control, and more effective tillage and water management, as well as by spells of favorable weather. These can offset reductions in yield for some years or decades, but cannot in themselves prevent a decrease in potential productivity. It takes a good deal of extra fertilization, specialized tillage, and the use of soil amendments (such as lime) to compensate for so great a loss of natural productivity as is due to the erosion of the topsoil.

In addition to the erosion of bared and cultivated soils by rainfall, a great deal of erosion can be caused by wind, particularly in dry areas of soils pulverized by modern machinery. One of the most spectacular and famous examples of wind erosion occurred in the 1930s in America's Great Plains region, popularly known as the Dust Bowl.[4] It resulted directly from the introduction of the plow into the vast, semiarid, wind-swept grasslands of the southern Great Plains. During most of the first three decades of this century, that region enjoyed a wet cycle. The prairie was green, and when plowed and planted it produced bountiful crops of wheat. But rainfall in that region, that includes parts of Colorado, New Mexico, Nebraska, Kansas, Oklahoma, and Texas, is inherently unstable. For example, in Oklahoma's Cimarron County, the average rainfall was about 500 mm annually between 1900 and 1930, and as high as 700 mm between 1914 and 1923; then in 1934 and the following few years it was barely 350 mm, and in some places it fell below 200.

It was mainly during the boom decades from 1900 to 1930 that farmers by the thousands settled the southern Great Plains region. They were oblivious to its hazards, and their farming methods were attuned to the rainfall regime of that wet period: they broke the sod, planted their seeds in the soft soil, watched the rain-satiated wheat come up and then harvested its bumper crop. In the Texas Panhandle alone, just south of Cimarron County, about 35,000 hectares were planted to wheat in 1909; by 1929 the area under wheat had expanded to 800,000 hectares. This enormous change in the land's surface was made possible by the mass production of tractors. Tractors later became the leading Dust Bowl villains. They were "snubnosed monsters," wrote John Steinbeck in *The Grapes of Wrath*, "raising the dust and sticking their snouts into it."[5]

Eventually the rain stopped, and was replaced by an inexorable dryness. 1934 was particularly hot and dry, and the land lay bare and parched under the merciless sun, its tillage-pulverized soil easy prey to the whipping winds. The towns, villages, and homesteads of the region clung to the denuded land. When the winds came, they swept up the loose soil as if it were talcum powder and rolled it along in billowing red-brown clouds that eclipsed the sun and obliterated fences and covered houses and choked animals and people. And so a great exodus began—40 percent of the people of Cimarron County, for example, abandoned their homes and moved out of the region during the 1930s.

They vanished as did their wheat crops of 1933, 1934, and 1935—planted there, but never grown.

The effects of the Dust Bowl were not confined to the region where it originated. The fine dust was wafted thousands of meters aloft and drifted across the entire continent. On May 12, 1934, the *New York Times* reported that the city "was obscured in a half-light similar to the light cast by the sun in a partial eclipse . . . and much of the dust seemed to have lodged itself in the eyes and throats of weeping and coughing New Yorkers." Washington, D.C., was overhung with a thick cloud of dust denser than ever seen before. The entire eastern seaboard of the United States was blanketed in a heavy fog composed of millions of tons of the rich topsoil swept up into the continental jet stream from the Great Plains. Even ships hundreds of miles out in the Atlantic found themselves showered with Great Plains dust.

The great American Dust Bowl of the 1930s is not merely a thing of the past. It is being reenacted on an equally vast scale in such regions as the Sahel in sub-Saharan Africa. The process by which a semiarid region is denuded of vegetation, and its soil destabilized to the point of uncontrolled erosion and degradation, is now called desertification.

The effects of prolonged cultivation, impelled by population pressure, can be seen in India, where intensive utilization of all arable land has resulted over time in marked loss of productivity. In the Deccan Plateau of Central India, there are millions of hectares of originally deep and naturally fertile soils derived from basalt, a volcanic rock, which have been under cultivation for many centuries, with the consequence that several meters of soil have been removed by erosion.

Land hunger resulting from the need for subsistence by rural populations leads to cultivation of marginal or unsuitable land. A notable example is the steady climb of cultivation onto increasingly steep mountain slopes. This is particularly true in the humid tropics, where alarming destruction is occurring because small-scale peasant farming is spreading onto land which cannot possibly sustain this use in the long run. Examples are the spread of cultivation up the mountain slopes in central Java, in the Machako and Aberdare districts of Kenya, the lower Himalayas of India and Nepal, and the Andes of South America. Tropical soils, originally endowed with organic matter that had accumulated under forest cover, can have their initial reserve of nutrients diminished in just a few years if the decomposition of organic matter is hastened by cultivation without replenishment. Added to that loss of organic matter is the loss of topsoil as sloping ground is exposed to intense rainfall; the result is a rapid reduction of fertility.

The Indonesian island of Java is an example of extreme overcrowding and erosion taking place on steep mountain slopes. The outer islands offer vast areas of undeveloped land, some of them with good soils and favorable rainfall. The government of Indonesia therefore conceived and attempted to implement a Transmigration Scheme aimed at reducing the excessive pressure on land in Java by transferring people to the outer islands. However laudable in its aims, this plan has been something less than a total success, and in some cases might have done more harm than good.

A colleague who worked for an international research institute provided a telling account of his visit to an outer-island development project, part of the Transmigration Scheme. He was startled to see monstrous bulldozers clearing a dense stand of trees by shaving off the entire topsoil and pushing it, along with the tangled mass of roots and trunks and branches, into the nearest stream. What remained was a smoothed parcel of land devoid not only of native plants but of topsoil as well—not a case of accelerated but of instantaneous erosion. My horrified colleague asked the foreman why he was removing the most fertile part of the soil, leaving only the acidic and nutrient-poor subsoil for the unfortunate would-be settlers to try to farm. The foreman only shrugged, saying that he was instructed to complete the job within a fixed schedule and was working as efficiently as he could.

Communal ownership of land, under which common property is often treated as if it is everybody's privilege to use but nobody's responsibility to protect, can encourage careless cultivation, over grazing, or hasty removal of firewood. The problem is worsened by the fact that soil conservation measures require an initial investment and subsequent maintenance costs, whereas contrary activities that avoid those costs produce short-term gains. So, where plans are made for conservation, there may be neither the will nor the funds to implement them. There is often conflict between the rural farmers' short-term objective of subsistence (as well as a government's objective of immediate national self-sufficiency) and the long-term objective of protecting a country's soil resources.

Another problem is deciding who benefits and who should bear the responsibility, particularly in the case of off-site damage. Especially difficult to quantify are the secondary effects of enhanced runoff and erosion resulting from soil denudation. Such effects include the necessity to clear roads, drain culverts, and dredge up the silt accumulating in channels and reservoirs. An example can be found in India, where overgrazing and overcultivation in the foothills of the Himalayas result in flooding and silting downstream, including disruption of irrigation and hydroelectric facilities. The peasant farmer in the hills may be the initiator of the problem, but he cannot afford to rectify the damage he has caused, nor to stop causing it. In this as in many other cases, the cost of corrective measures can only be undertaken by the state, in whose long-term interest it is that the soil be conserved, and that cost must include providing the poor hill farmers with an alternative livelihood that is ecologically sound.

An extreme example of the off-site damage caused by enhanced runoff and erosion is seen in Bangladesh. Every year late August brings floods: not in raging torrents but in the slow rise of the Ganges, the Brahmaputra, and the Meghna—three of the world's most copious rivers. In 1988, nearly three quarters of the country was submerged, thousands of people perished, and an economy already one of the world's poorest was devastated once more, as it had been repeatedly in past years. Containing the floods of Bangladesh is technically difficult and economically onerous, and cannot be done within Bangladesh alone. The waters that flood Bangladesh come mainly from India and partly from Nepal, so any fundamental solution to the problem must be regional, involving the countries that control the watersheds as well as the country that receives the floodwaters. Assessing responsibilities among these countries, and the costs of the damage and its mitigation, calls for the wisdom—and the authority—of a Solomon.

It is naive to think that conservation is such a good thing that as soon as the people are shown what to do, they will immediately and enthusiastically begin to do it. Even in the United States, for example, where the technical know-how and the government-sponsored technical and financial assistance are both available,[6] the implementation of conservation programs is far from universal. It is difficult to use conventional accounting procedures to justify soil-conservation works, because one cannot very well assess in monetary terms the future value of the soil that is saved today, or the diminished value of the soil that will be eroded otherwise. For a government to promote conservation effectively, it may need to use both the carrot and the stick, the former including persuasion and subsidies for adopting conservation, the latter consisting of penalties for infractions.

Agriculture need not, in itself, cause degradation of the land. Not every farming practice necessarily changes the soil for the worse. Through experience and research, practices have been established that can maintain a stable system of production.[7] A package of physical, chemical, and biological treatments known as conservation management can in fact enhance, rather than diminish, the potential productivity of soils. This package includes controlled drainage, organic fertilization by means of manures and composts, mulching, non-compacting and non-pulverizing tillage, avoidance of overland traffic, use of soil amendments such as lime and gypsum, and the growing of soil-building plants such as legumes (including their incorporation into the soil as "green manures"). Lands which are too steep or erodible to be so protected and enhanced should properly be retired from cultivation, and the people trying to eke out their existence from it should be given alternative and better opportunities.

The U.S. Soil Conservation Service has developed a multi-component program to protect arable land against water erosion. As a first line of defense, a diversion drain can be built at the upper edge of the parcel to intercept any surface runoff that might otherwise flow from higher ground onto the arable land. The second component is a series of graded channel terraces dividing the arable land into contoured strips, each with a channel built on a gentle grade so that runoff from the arable land can be drained away at a non-erosive velocity. The third component is a grassed waterway that can collect the runoff discharged from the graded channel terraces and lead it safely to a natural watercourse. This system is effective only on moderate slopes of up to about 8 degrees—13 to 15 percent, as reckoned by the engineers. On very gentle slopes (up to 3 percent), it is often possible to control erosion by dividing the land into contour strips separated by bands of grass, and by carrying out all cultivations on the contour rather than up and down the slope (as is done by careless cultivators who thereby enhance, rather than retard, runoff and erosion). On steeper slopes, some form of bench terracing is generally employed to convert the slope into a series of steps with nearly horizontal ledges, and nearly vertical walls between them. However, the most effective way to prevent erosion from the start is to avoid the sort of tillage that pulverizes the soil excessively and bares the surface to the action of wind and rain.

Wind erosion is usually controlled by maintaining the topsoil in a rough, cloddy, and if possible also moist condition, and by leaving a stubble mulch of plant residues to cover the soil surface. Windbreaks, which are obstructions established along lines perpendicular to the direction of the prevailing winds, can serve to reduce the speed and turbulence of wind near the ground. Such windbreaks can be mechanical (fences or screens), or—more often—vegetative. Rows of trees and shrubs are commonly used, because in addition to reducing soil erosion they can produce food, browse, firewood, and wildlife refuges. Series of such vegetative windbreaks are called shelterbelts. Their effect on reducing wind erosion depends on their height, density, and spacing. Generally, the sheltered distance on the downwind side of a windbreak is about 10 to 20 times the height of the windbreak. Somewhat porous windbreaks have been found to be more effective than solid walls, as they reduce wind turbulence on the downwind side of the windbreak. However, leaving gaps might channel the wind and cause localized erosion. Shelterbelts also have disadvantages. They occupy land, and reduce the growth of crop plants immediately adjacent to them because they shade the crop and compete with it for water.

Despite the availability of such soil conservation measures, in practice most agricultural activities in the field do accelerate erosion to some degree. However desirable, the total or absolute elimination of erosion, in many cases, may be an impractical goal, as the cost of achieving it is likely to be too expensive. A more practical goal is to limit soil erosion to a level that is tolerable—a rate of soil loss that is roughly equal to the rate of topsoil regeneration, aided by good cultural practices, that can be sustained without significant reduction of potential productivity. If, for example, the farmer can regenerate 1 kilogram of topsoil per square meter per year (equivalent to about 0.8 millimeter of soil depth per year or 1 centimeter in about 12 years) then the tolerable annual soil loss is 10 tons to the hectare. The achievement of even this limited goal, however beneficial to the environment and to society, may be much more expensive than the readily discernible benefit to the farmer himself. Any farmer conscientious enough to practice strict soil conservation may suffer from the competition of farmers who avoid doing so and are therefore better able (in the short run of some years or decades) to produce food at lower cost and sell at lower prices. Hence farmers are often hard-pressed to finance soil conservation without some form of subsidy.

In the early 1960s, the U.S. Soil Conservation Service's guidelines for soil loss tolerance set a maximum of 5 tons per acre (12.6 tons per hectare) per year. In actual practice, soil losses 10 times as great are common throughout the U.S. In many other countries, rates of soil loss are just as great, and in tropical regions with intense rainstorms and sloping terrain the loss rate can be even greater. The task of combating soil erosion is a continual challenge to agriculturists the world over.

I hear, and I forget;
I see, and I remember;
I do, and I understand

KUNG FU-TSE

21

THE "SORROW OF CHINA"

THE HIGHLANDS REGION of Northwest China is the world's most extensive and massive deposit of loess—a yellowish, soft, silty sediment swept up by the continental winds from the great Central Asian desert belt and laid down in a region far removed from its origin.[1] Deposition of loess in China apparently began in the early Pleistocene era, some 1.2 million years ago. Over the millennia, the loess has accumulated to form a thick blanket of uniform soil material—indeed the deepest soil in the world. It can be highly productive when properly cultivated, but is by nature (being loosely laid desert dust) extremely vulnerable to subsequent erosion. When pulverized in the dry state, it reverts to a fluffy powder that is easy prey to the swirling air currents; and when pelted by raindrops or lapped by running water, it slakes down to form a muddy suspension all too ready to be carried away into the region's streams and rivers.

Deposits of loess occur in Central and Eastern Europe (most notably in the Ukraine), the great plains of North America (the United States and Canada), in parts of South America, and in the Middle East. Nowhere, however, are the deposits of this distinctive soil material as massive as in China, where they

169

cover a region of many thousands of square kilometers, in beds that are hundreds of meters deep.

Loess is an astonishingly homogeneous soil material, devoid of stones and with very little organic matter. Although not mutually cemented by humus, the jagged particles tend to pack together and interlock so as to impart to the soil an unusual capacity to maintain vertical walls without slumping, as long as the material remains dry. When saturated with water, however, masses of loess lose their cohesiveness entirely and may collapse spontaneously. On China's northern plateau one sees great bluffs of loess, where the plateau is dissected by rivers. The loessial material forms high vertical cliffs, into which people have, since ancient times, dug caves for dwelling and for the storage of grain. When undercut by scouring streams, huge chunks of these cliffs come crashing down abruptly and may block or divert an entire river.

The great river which drains the loessial plateau of China is called the Yellow River, and it derives its name from the color of the sediment that it carries off the plateau as it follows its turbulent course toward the plain in the east. So laden is the Yellow River with loessial silt that it is considered to be the muddiest major river in the world. On average, its churning and frothing water contains an astonishing 34 percent by mass of silty sediment, and appears to be a rippling tide of liquefied mud, resembling thick lentil soup. Along its headwaters, this voracious river, with its tributaries, continues to gnaw away at the land, whereas farther downstream the river slows down and deposits sediment until its bed rises high above the level of the adjacent flood plain.

From time to time, as torrential rains visit its catchment, the river surges, overflows its banks, bursts the confining levees built so laboriously by hand, and inundates huge tracts of land, destroying countless villages and causing human displacement and suffering on an enormous scale. It is an ever-present, ever threatening fact of life in China that the same benevolent river that serves as the source of life for millions can suddenly decimate those who depend on it. No wonder it has been known for centuries as "The Sorrow of China." When the river finally

reaches its estuary and spills its remaining load into the sea,
that body of water is so tinted by the pale golden loess that it
is called the Yellow Sea.

The Chinese have been trying to contain, moderate, and
harness the Yellow River by building flood-control dams—which
also serve for the generation of hydroelectric power—along the
upper reaches of the river, and by reinforcing the levees along
its lower reaches. The agricultural stability of the North China
Plain is intimately associated with the loessial upland catchment
from whence the Yellow River draws its water and sediment.
The catastrophic rate of soil erosion in the catchment area is
the source of the most dangerous threat to the North China
Plain. The inherent erodibility of the loessial soil is magnified
by its tendency to form a surface seal, or crust, which inhibits
infiltration and induces greater runoff.

The Sisyphean effort to control runoff and erosion from the
loessial plateau involves the construction of terraces on the steep
slopes. This effort is facilitated by the unique capacity of the
loess to sustain vertical cuts, but is thwarted by its extreme
erodibility. Most of the terraces are built by hand. The surface
of each plot is leveled, and the edge of each is marked by an
earthen embankment with a height of about 20 cm above the
level of the plot. That height is considered sufficient to impound
the water during heavy rains so as to ensure complete infiltration
and prevent the escape of runoff. From time to time, however,
torrential rains scour the embankments of these terraces, as they
did in the summer of 1978. Occasionally, cracks form in the
vertical walls between the step-terraces, causing the collapse of
huge blocks of earth. So the task of combating erosion in these
circumstances is a never-ending, arduous effort. Mechanical or
engineering means must be complemented with the use of dense
vegetative coverage to protect the sloping sections of the erodible
soil. To date, however, these efforts have resulted only in partial
success. The ancient, vicious, capricious dragon yet lurks in
the incompletely tamed river, which continues to wear away
the land—and then tumbles on its way, seemingly as yellow,
as boisterous, and as contemptuous of human efforts as ever.

It is along the lower reaches of its course that the Yellow

River does its greatest damage. The North China Plain, lying east of the loessial plateau, holds a key position in that country. It is the largest plain in China, covering an area of 300,000 square kilometers. The soil cover is the gift of the numerous streams descending from the highlands. The surface material is redeposited loess with a silty loam texture. The region has a long geological history of subsidence and deposition, so the unconsolidated deposits are very thick, ranging from less than 200 meters to more than 1,300. The entire region is underlaid by a series of groundwater reservoirs that consist mostly of fresh water but in some places are highly saline. The region has a temperate climate, with a warm summer and a cool winter. Parts of the region are subhumid, with a mean annual precipitation of more than 800 mm, but other parts are semiarid with no more than 450 mm. The precipitation varies by season and over time, with a marked concentration of rainfall in summer and great variability from year to year.

The Yellow River discharges an average 1.6 billion tons of silt each year at the point where it emerges onto the plain. The main part of the plain is in fact a huge alluvial fan built of the colossal quantity of silt deposited by the shifting river. As is usual with heavily silt-laden rivers, the Yellow has raised its bed high above the surrounding alluvial plain and it flows between natural levees. To prevent the river from rising above its levees and to keep it from changing its course capriciously, humans began—as early as 2,500 years ago—to reinforce the natural levees by erecting dikes and embankments. Unfortunately, this expediency, requiring enormous investments of labor and time, has proven again and again to be ultimately self-defeating. When the moving sediments, once spread far and wide, are artificially confined within the embankments, such sediments tend to settle to the bottom as the river slows down in the flat plain and the water loses its capacity to carry particles in suspension. The cross-sectional area of the channel is thus continually reduced, and its capacity to transport water as well as sediments is diminished. Consequently, the river becomes more prone to rising and spilling over its banks following episodes of heavy rains occurring on its watershed. The result is a much greater potential for disastrous flooding of the plain.[2]

Historical records indicate that more than 1,500 dike breaches occurred over the two thousand years preceding 1950, and that these breaches often produced devastating floods. Moreover, the river has long had a disconcerting habit of changing its course about once in a century, and the area affected by these unpredictable events has included three-quarters of the North China Plain. The scars of more than 15 ancient courses are still visible. The present channel as it runs through the plain has a length of 818 kilometers, the upper section of which has apparently remained in the present course since 1376 c.e., whereas the lower section dates back to only 1855. Over much of its length, the channel is about 10 meters above ground level, so the river actually flows on a high ridge overlooking the plain, with all its villages and towns constantly under threat of flooding. To keep the river in check, the Chinese are forced to raise the dikes by one meter each decade. The need to reinforce and raise these dikes repeatedly, over a length totaling 1,400 kilometers, is a constant strain on the Chinese economy, but still does not solve the basic problem of this unruly river.

Agriculture in this region has a long history. In ancient times, the development of water management methods transformed this area into the cradle of the brilliant civilization of China. In fact, the entire history of the Chinese agricultural economy is linked to the control and utilization of water resources, which apparently started in this region as early as 2,000 b.c.e. The diversion of water from the Yellow River to irrigate paddy fields started early in the Han Dynasty. Some famous 1,000– 2,000-year-old canals are still in use today.

The land is cultivated with meticulous care. The flat terrain and the deep and fertile soil material account for the fact that over 18 million hectares (fully 60 percent of the entire region) have been under the plow by the "Farmers of Forty Centuries."[3] This land area represents more than 17.5 percent of all the cultivated land in China. Although the cultivated land area is less than one-tenth of a hectare per capita, there is little if any room for expansion of cultivation. So the demand for food and other agricultural products has exerted a great pressure on the land available.

Although the Yellow River has been kept confined within its dikes for several decades, the hazard of inundation in fact is increasing, as the channel's capacity to carry water is diminishing. Despite the immense effort devoted to their reinforcement, the dikes are now more vulnerable to destruction by rising waters, and so is the adjacent plain. The success of flood prevention during the last few decades provides no guarantee for the years to come. Historical records show that peak flow rates exceeding 30,000 cubic meters per second occurred in previous centuries and hence are bound to recur sooner or later. In contrast, the highest peak flood discharge in recent years was that of 1982, and was only half the historical peak flow rate.

Another problem in parts of the North China Plain is the sinking level of water in the aquifer, caused by over-pumping from tubewells providing irrigation and domestic water. Unfortunately, as much water is wasted in irrigation here as it is in many other countries. More than 50 percent of the water extracted from the Yellow River is lost while passing through unlined canals, and such extraction impairs the river's vital capacity to transport sediment. Apart from the sheer waste involved in seepage from unlined canals, there is the danger of raising the water table excessively, thereby causing waterlogging and salinization.

In China as elsewhere, much can and must be done to promote higher efficiencies in the conveyance, application, and utilization of water in the field.

*As the fire that burneth the forest, and
the flame that setteth the mountains
ablaze.*

PSALMS 83:15

22

DEFORESTING THE
EARTH

FOREST ECOSYSTEMS PLAY a crucial role in creating and maintaining the earth's life-support system. Acting as the lungs of our terrestrial environment, forests assimilate carbon dioxide and release the elemental oxygen vital for the respiration of all animals, including humans. They protect the soil from the direct impact of rainstorms and from erosion by water and wind. Forests moderate the local microclimate, and, where they cover extensive areas, the regional macroclimate as well. With their accumulated organic matter—living biomass and its dead residues—they serve as dynamic sponges, absorbing and then gradually recycling water and nutrients, thus regulating their flows. Forests also absorb, filter, and purify potentially toxic gases and aerosols (the latter being minute particles, such as dust, that are borne in suspension by the atmosphere). Finally, forests constitute the ideal common habitat for an enormous variety of plants and animals, a treasure-trove of biological diversity, the luxuriant and exuberant culmination of our biosphere's long evolution.

Before humans began to interfere with the natural habitat of the forest and to modify and then decimate it, the continents were carpeted with more than six billion hectares of woods and

shrubs. Over time, such activities as land-clearing for crop production, gathering of fuelwood, commercial harvesting of timber,
and livestock husbandry, together have caused the earth's forests
to shrink to some four billion hectares—a third less than existed
originally. The shrinking of the forests deprived numerous species
of plants and animals of their habitat, indeed of their existence.
However, for centuries this reduction of the earth's biological
stock did not seem to hinder human material progress. In fact,
the clearing of trees to expand food production, and the harvesting
of forest products, were important aspects of economic and social
development. In recent years, however, the relentless loss of
tree cover has begun to impinge directly on the economic and
environmental health of numerous nations, mostly in the Third
World. With the unrestrained march of deforestation, the ecological integrity of many areas is disintegrating, causing severe
soil loss, aggravating droughts and floods, and permanently reducing land productivity.

Man's attack upon the forest began very early in pre-history,
during the hunter-gatherer phase which long preceded the dawn
of what we commonly call civilization, and it has continued
and intensified following the advent of agriculture and ever since.[1]
Fire was the principal means for clearing the thicket. Its use
enabled early man to open up the landscape so as to facilitate
hunting, grazing, and eventually cultivation.

The dangers to natural flora and fauna inherent in the unrestrained use of fire were recognized in antiquity, as can be seen,
for example in the edict of the Indian emperor Ashoka, who
lived in the third century B.C.E.: "Forest fires should not be lit
unnecessarily and in order to kill living beings." In the book
of Judges, the parable of Jotham includes the metaphor: "Let
fire come out of the bramble and devour the cedars of Lebanon."
The prophet Jeremiah warned Judea that the Lord might "kindle
a fire in her forest and it shall devour all her environs." And
the prophet Joel lamented: "Unto Thee, O Lord, do I cry, for
the fire hath devoured the pastures of the wilderness and the
flame hath set ablaze all the trees of the field; yea, the beasts
of the field pant unto Thee, for the water brooks are dried
up."

Especially vulnerable to fire were the forests of the Near East, and of all the lands surrounding the Mediterranean. Here, half the year is hot and totally dry, and periodically subject to the searing winds blowing in from the nearby deserts, so the parched and highly combustible undergrowth of grasses and shrubs becomes easy prey to the smallest spark, bursting instantly into flames. The resulting conflagration can quickly engulf an entire forest and reduce it to ashes in a matter of hours. Occasional fires are due to spontaneous natural causes, such as lightning, so in fact the occurrence of such fires is probably intrinsic to the region's ecology. However, humans have greatly increased the frequency of the fires, and followed them up with equally destructive activities, including felling trees, trampling and over-grazing the land, tilling the soil, and reshaping its surface.

In the wake of a fire, most of the indigenous grasses can regenerate during the next rainy season, but not before the temporarily denuded soil has borne the brunt of the erosive early rains. Many tree species are sensitive to repeated fires, however, and do not regenerate readily, especially if their remnants are cut down and their saplings are grazed excessively, or if the soil is cultivated. As a consequence of human activity, the dense climax vegetation of evergreen and deciduous trees which once covered the hills and valleys of the Mediterranean littoral has been supplanted by an open patchy landscape of grasses and shrubs with widely spaced trees and disjointed fields. The eradication of forests was accelerated by the cutting of wood for use as fuel, and as timber for houses, temples, weapons of war, and ships. The consequence of all this activity was greater than just the loss of the tree cover and the forest habitat. It also included the permanent loss of the fertile soil substrate. Not much has remained of the original luxuriant forests of the Mediterranean region, nor much of the soil that long ago covered the hillsides. The wound inflicted on the earth can be seen in every country around the Mediterranean,[2] and it is exceedingly sad: skeletal, stony, shallow soils occurring as pockets amid the protruding outcroppings of bedrock.

Only a few areas in the region are still forested, yet even these meager remnants are still being put to the torch. In the summer of 1989, Spain, for example, experienced its worst season of forest fires, exacerbated by a hot and parched summer. In the first nine months of that year, 11,420 fires destroyed 220,000 hectares. A similar fate has befallen other parts of the region, including Portugal, Sardinia, the French Riviera, Corsica, Turkey, and Greece. It appears that only a small percentage of these fires can be attributed to natural causes. Rather, the principal cause seems to be the persistence of obsolete and dangerous agricultural practices like the burning of brushland and pastures at winter's end, when the fires are fanned by strong winds. Lit cigarettes, bonfires, and deliberate arson are also important causes.

The effects of continued and widespread fire-abuse are extremely serious. Fire-ravaged soil, stripped of all undergrowth and litter, is less able to retain water during heavy rains, so runoff and erosion ensue, leading to violent flooding of valleys and silting of streams during the rainy season, and to diminished flow rates during the dry season. This increased variability in stream flow makes water management and regulation a much more difficult and expensive task.

The Mediterranean basin is only the first of the earth's regions to undergo deforestation. Beginning in the sixteenth century, the expanding agricultural and industrial needs of the Renaissance societies spurred the clearing of large tracts of forest in Western Europe.[3] Both the French and the English so depleted their domestic forests that they were forced by the mid-seventeenth century to import ship timber from distant countries to support their expanding maritime ventures. France, once 80 percent forested, remained with trees covering only 14 percent of its territory by 1789. A similar fate befell the United States of America. When the Pilgrims arrived, circa 1630, the forest cover in what is now the contiguous 48 states totaled some 385 million hectares. As colonization spread, forests dwindled. By 1920, trees covered less than 250 million hectares, a third less than when European settlement began. As of 1982, the forest cover of the contiguous states had diminished to 233 million hectares.

Agencies of the United Nations—the Food and Agriculture

have recently attempted to inventory the current status of the
global forest resources base. They give the following estimates:
Closed forests, where the tree canopies completely shade the
ground, cover some 2.8 billion hectares worldwide. Another
1.3 billion hectares are open woodlands, including the savannas
of Africa and the cerrado in Brazil. Collectively, then, forested
lands cover approximately 4.1 billion hectares, an area almost
triple that in crops. Shrublands and forest regrowth on temporarily
abandoned cropland bring the total area supporting woody vegetation to about 5.2 billion hectares, almost 40 percent of the
world's land. However, the most worrisome finding of the U.N.
assessment is that tropical trees are being cut much faster than
natural regeneration or artificial reforestation are replacing them.
For tropical areas as a whole, 11.3 million hectares were cleared
annually in the early eighties, while only 1.1 million were being
replanted. Thus 10 hectares were being cleared for every 1 planted.
In Africa, the ratio was 29 to 1, and in Asia 5 to 1.

Overall, the conversion of forest to cropland is by far the
leading direct cause of deforestation in many tropical regions.[4]
Population growth, inequitable land distribution, and expansion
of export agriculture have greatly reduced the area of cropland
available for subsistence farming, forcing many peasants to clear
virgin forest to grow food. These displaced cultivators often
practice continuous cropping methods that are ill suited to tropical
forest soils, which are extremely fragile. Cultivation leads to a
dramatic increase in soil loss, from perhaps 1 ton per hectare
annually to about 50 tons, and in places to over 100. Within
a few years, the soils become so depleted of nutrients, and so
eroded, that peasant colonists are forced to clear more forest
tracts in order to survive.

In the time-honored practice of shifting cultivation, indigenous cultivators in the tropical regions of Africa, Southeast Asia,
and the Americas had for countless generations cleared new fields
every few years by the traditional slash-and-burn method, then
allowed forest regrowth to restore soil fertility before they returned
to clear and plant crops again many years later. The FAO has

estimated that shifting cultivation is responsible for 70 percent of closed forest clearing in tropical Africa, nearly 50 percent in tropical Asia, and 35 percent in tropical America. But this once-sustainable practice is being undermined as population pressures force cultivators to return to the same plots earlier and earlier, before these plots have had the time needed for complete restoration of soil fertility. Thus, intensified shifting cultivation can lead to serious land degradation. A vicious cycle then begins: soil depletion and erosion delay forest regrowth, so the period needed for recovery grows longer even while the period allowed for recovery gets shorter. The process of degradation accelerates and in time makes the soil unproductive. Indonesia, for example, has more than 16 million hectares of nonproductive land that is incapable of supporting agriculture or forest without major rehabilitation.

An additional agent of destruction is the collection of fuelwood for domestic purposes. More than two-thirds of all the people in the Third World rely on wood for household cooking and heating. Rural people depend on wood almost entirely, even in oil-rich Nigeria. The practice of collecting fuelwood is particularly unsustainable in the semiarid woodlands of Africa, where the population density is high and the incidence of drought (often coupled with the practice of overgrazing) reduces the growth rate of woody vegetation. The degree of denudation is most extreme in the areas surrounding large cities in Asia and Africa, as the concentrated demand there overtaxes available tree stock. Recent data obtained from remote sensing by satellite show that, in less than a decade, the forest cover within 100 kilometers of India's major cities dropped by 15 percent or more; the area surrounding Delhi lost a staggering 60 percent. Attempts to promote the production of fast-growing trees by farmers as a regular component of their farming activities (in a system called agro-forestry) are promising, but not yet widely adopted.

Deforestation on the largest scale is taking place in Brazil,[5] where, until very recently, it was official policy to encourage land clearing in Amazonia for the sake of development. The story is instructive enough to be told in some detail. In 1970, the Brazilian government devised a plan to solve two pressing national problems at once: squalid conditions for peasants in northeast Brazil, and concern about the nation's sovereignty in the largely uninhabited western and northern provinces. It then began construction of the Transamazon Highway in hopes of transferring a million landless people into the western frontier and settling them there, in what seemed to be a bold and brilliant program of national development. But the program faltered at first. Fewer than 20,000 people settled there in the first years. The land simply would not support the monocultures the government was advocating. Slash-and-burn clearing gives the soil a temporary boost from the nutrients formerly bound in the vegetation, but those nutrients are soon used up or leached. Most of the immigrant farmers found that after one or two good years, crops failed. They abandoned the depleted land and moved back east, even more impoverished than before.

Still determined to advance the program, the government then instituted tax incentives for cattle ranching, lured in large measure by the insatiable U.S. appetite for imported beef—a purpose for which the rain forest soils are as unsuited as they are for most types of monoculture cropping. Most of the commonly sown pasture grasses fare poorly on the infertile tropical soils, especially as the nights are warm—requiring high rates of respiration that use up the plants' reserves of carbohydrates—and as there is no cold season to control pests. And so a destructive pattern was set in motion, impelled by poverty and greed and short-sighted policies: Peasants cleared the land, failed within a few years, and turned the property over to large cattle operations, which got by for a few years on tax breaks and land speculation. When the land was played out even for the cattle ranchers, they too moved on to new lands. The real acceleration of Brazilian deforestation began in the mid-1970s, when the roads reached Rondonia, where soils for food crops are at least marginal. Since 1968, Rondonia's population has grown by 1,000 percent. Movement to the western states now amounts to a half-million people per year.

Some environmental activists believe that the Brazilian government should be pressured into taking drastic action to halt the arsonists and colonizers of its rain forests. However, coercive decrees are not likely to relieve the growing problem of the dispossessed poor—the major cause of loss of the tropical rain forest. A far better approach is to improve the productivity of cultivated land so as to relieve the pressure to clear ever-new land. Some of the techniques are well known: crop rotation, intercropping, drip irrigation, judicious use of fertilizers and chemical amendments, and integrated pest control. Nothing is likely to work, however, if the nation's explosive population growth is not moderated.

Much can be learned from the indigenous people about sustained harvesting of the natural products of the rain forest. Recent investigations have revealed the existence of an ancient Amazonian civilization that prospered for many centuries in the very areas that are now being destroyed by modern "development." The native inhabitants of the Amazon evolved a sophisticated blend of agriculture and forestry that utilized the region's resources without destroying them. Many lived in large settlements atop earthen mounds, protected from the floods. They relied on the natural products of the forest (including nuts and resins), as well as on agricultural crops.

The Amazonians' farming methods were based on periodically clearing a circular field, by felling several large trees at key junctures so their crowns fell on the periphery of the circle. Under carefully controlled conditions, without endangering the surrounding forest, they would then burn the fallen brush so that rains could wash the fresh ash into the soil. Into each new field, they would plant crops in concentric rings. Included were yams, sweet potatoes, manioc, maize (corn), rice, beans, and cotton, as well as such perennial fruit crops as papayas, bananas, pineapples, and mangoes. After 10 to 15 years, as the yields began to diminish, they allowed the forest to re-cover the field. The succession of native plants re-invading the clearing then provided a secondary harvest of medicinal herbs and other useful plants and animals. In some places, the re-establishment of the forest was managed by the mixed planting of selected native

trees that provided edible fruits or nuts, building materials,
and a variety of products such as pesticides, medicines, cosmetics,
and fragrances. The abandoned fields were allowed 50 years or
more to recover productivity before being cleared again.

The productivity and ecological soundness of these traditional systems was unappreciated until recently, if only because official economists and analysts tend to prefer market-based, cash-producing economic activities to subsistence economies. However, it now appears that the traditional approach can be adapted so as to supply a broad range of marketable and exportable products while still leaving the forest cover basically intact. The sustainable harvesting of these products can be more profitable than the widespread destructive logging and cattle ranching still practiced today.

Of late, the government of Brazil has begun to discourage forest clearing for ranching, and the rate of deforestation seems to have slowed down. Still, deforestation is so extensive that some scientists believe it may affect the regional climate. (Satellite imagery has shown significant portions of Amazonia to be under a pall of smoke from the extensive burning of the forest.) Evidence from the Amazon basin suggests that the tropical forest there returns as much as 75 percent of the rainfall to the atmosphere through evaporation and transpiration, forming new rain clouds. The overall disposition of rainwater includes 25 percent that flows into streams and rivers as runoff, another 25 percent that is readily evaporated from the raindrops remaining on the leaves of the multi-storied forest canopy, and a final 50 that is transpired subsequently from that foliage. A forested area was found to collect and return to the air 10 times as much moisture as a barren area, and twice as much as shrubland and grassland. Thus the Amazonian forest actually recycles most of its moisture through its enormous evaporative and transpirative surface, and in the inland areas only a small fraction of the rainfall comes directly from the ocean. The ability of such an extensive forest to create its own clouds and rainfall may thus affect the regional climate significantly, and may even have a bearing upon the global weather pattern.

Incidentally, human activities in the tropics may have an

important effect on atmospheric chemistry. The burning of trees and decomposition of soil organic matter, which occur when virgin forest soil is cleared and cultivated, result in the oxidation of carbon and its release into the atmosphere as carbon dioxide (CO_2). The annual emission of CO_2 from cleared tropical forests has been estimated to be as much as 20 percent of the total global emission of this gas, most of which is due to the burning of fossil fuels. In addition to CO_2, the conversion of land into rice fields and cattle pastures also increases emissions of methane, another gas that is radiatively active and enhances the now-famous greenhouse effect.[6]

Logging is responsible for a quarter of the destruction in the Brazilian rain forests, and is the main cause of similar destruction in Southeast Asia. Indonesia, Papua-New Guinea, Burma, and the Philippines have cut more than half of their primary rain forest and shipped the valuable timber to Japan and to Western nations.

Although the deforestation rate has recently declined in some areas, it is doubtful that the forest cover in the Third World will stabilize anytime soon. Unfortunately, the forces that drive deforestation are more compelling than ever, and replanting efforts are still woefully inadequate to stem the progressive loss of tree cover.

In some temperate regions, by contrast, pressures on forests have lessened substantially following several centuries of clearing for agriculture and urban development. Forest cover in many European countries has been fairly stable, and in some it has even been increasing as marginal agricultural land is allowed to revert to woodland, and as conscious efforts are made to plant trees. Both Britain and France have been gradually increasing their forest cover; in France the fractional area under forest has risen from its historic low of 14 percent to about 25 percent of the land area.

A new scourge now threatens the forests of the industrialized countries in the form of air pollution, and more specifically of acid rain resulting from industrial emissions of sulfur and nitrogen oxides. These oxides dissolve in the rain and make it strongly acidic. Consequently the slight natural acidity of rain, due to dissolved carbon dioxide, has been increased a hundredfold in some areas (from an original value of 5.6 to less than 3.5 on the logarithmic pH scale). Acid rain has corrosive effects upon buildings and monuments, especially those made of marble, limestone, and metals. It also affects soil and bodies of surface water. Some streams and lakes have already become inhospitable to many species of indigenous fish. Acid rain also attacks trees, and has thus far caused the deterioration of some 31 million hectares of forests in Central and Northern Europe. The same scourge now affects large area of forests in the eastern parts of the United States and Canada.[7]

We are reminded of the H. G. Wells adage that the history of humanity (and consequently the history of the earth's environment subject to human meddling) is a race between learning and disaster. Repeatedly, human action designed to alleviate one disaster ends up bringing on another, as when efforts to eliminate hunger by expanding agriculture result in deforestation and erosion. The race is never completely won, especially as we are waging it against our own short-sighted impulsiveness.

23

MAN-MADE DESERTS

THE WORLD'S ARID REGIONS are undergoing a process of land degradation commonly described as desertification.[1] The people of these regions are engaged in a life-and-death struggle to protect their threatened lands, their source of subsistence, and their cultural traditions. Their extraordinary plight has drawn the special attention and concern of people everywhere.

The term "desertification" was originally coined by French scientists LeHouerou to describe the northward advance of the Sahara in Tunisia and Algeria. It entered popular parlance in the mid-1970s, and has been in vogue ever since. To many, this term conjures up the monstrous image of a relentlessly advancing desert, a tidal wave of shifting sand encroaching upon and devouring fertile farmland. *Deserts on the March, The Sand Swallows Our Land,* and *Rolling Back the Desert* are titles of books and articles that reinforce that specter. The emotive word and the associated imagery have given impetus to a decade-long effort by the United Nations Environmental Programme (UNEP) to combat the spread of deserts. Countries located on the southern fringes of the Sahara have been granted international aid to plant millions of trees as a barrier, presumed to be an impregnable

Maginot Line, to hold off the invading sands. But the image might be a mirage and the effort misdirected.

To be sure, there is a very real problem, but the perception and diagnosis of it may have been wrong. Desert-like conditions are actually forming in the arid and semiarid regions of North America, sub-Saharan and Southern Africa, the Near East, and Southern and Central Asia. Although various forms of land degradation are occurring throughout the world, the damage is greatest in these dry regions, which cover roughly one-third of our land surface and are home to some 850 million people—over one-sixth of the world's population. But attempts to encapsulate the problem in the catch-all term "desertification" have obscured its true character and confused the search for its amelioration.

The UNEP defines desertification as "the diminution or destruction of the biological potential of land that can lead ultimately to desert-like conditions." That definition leaves open several questions, such as the degree of damage to the vegetation and soil that is considered "destruction" (presumably non-recoverable), and the exact meaning of "desert-like conditions." It also begs the question as to what might cause that destruction of an area's biological potential. Is desertification due to a climatological process of increasing aridity that is independent of human action? Or, is it due entirely to human action even without any fundamental change in climate at all? Or, finally, might there be an interactive combination at play, in which human intervention and climate change feed upon each other, causing habitat degradation?

Ever since the U.N. Conference on Desertification held in Nairobi, Kenya, in 1977, the issue has become a *cause célèbre*, and consequently a *raison d'être* for organizations and bureaucracies devoted to its alleviation. Popular reports have tended to exaggerate the problem, and many unfounded assertions and hyperboles have gained currency. Among the most frequently repeated is the claim that the Sahara desert has expanded southward by about 100 kilometers between 1958 and 1975, an average of about 6 kilometers per year. The president of the World Bank stated in 1987: "We know we must stop the advance of the desert. . . . In Mali, the Sahara has been drawn 350 kilometers south by desertification over the past 20 years." Another assertion made by the UNEP is that "currently 35 percent of the world's land surface is at risk. . . . Each year, an area of 21 million hectares is reduced to near or complete uselessness." Such figures have been enshrined as institutional fact, cited and recited so frequently as to become articles of faith. They have focused attention on the edges of existing deserts, as if these are the front lines in a war against the desert's invasion, and therein lies the major fallacy.

Unfortunately, there is precious little scientific evidence regarding the expansion of the desert. Nor is there sufficient information regarding the extent, rate, or degree of land degradation in the regions adjacent to the desert. The original delineation of the Sahara's edge made in 1958 was based on very limited data. Moreover, in 1975 there was a drought, and so the frequently cited comparison of 1958 with 1975 failed to distinguish between the temporary effects of the drought and any possibly permanent changes in the boundaries of the desert.

There are, of course, places where the desert can be seen encroaching on neighboring land, and where it can be held back, more or less literally. By far the larger problem, however, is not the advance of the desert along its edge, but the deterioration of the land due to human abuse in regions outside—and in some cases quite removed from—the desert itself. The problem is emanating not from the desert but from the centers of population; not from the spread of the sand dunes but from the spread of people and their mismanagement of the land. Therefore, protecting the front line may do nothing to halt the degradation behind it. The true challenge is not so much to stop the desert at the edge of a region, as to protect the region itself from internal abuse of its vegetation and of its soil and water resources.

Correct diagnosis is essential if the problem is to be redressed. If, for example, the problem is believed to be one of short-term drought, then stop-gap food aid should be enough to maintain the people on the land until the rains return and they can become self-supporting once again. If, however, there are already

more people in the area than the land can sustain, the provision
of relief will merely allow the population to continue growing,
and thus defer the problem to a later date, at which time it
may be worse. If the perception is that the climate is changing,
then permanent withdrawal of the population or a drastic change
in their mode of subsistence might be indicated. In any case,
treating the edge of the unpopulated desert will do little to
help the populated heartland that is bearing the brunt of the
drought. There is no reason to plant a belt of trees in order to
halt the progress of the desert if it is not moving, or, on the
other hand, if the desiccating trend will prevent the trees from
growing.

The debate over desertification has tended to focus upon a
particular region of Africa south of the Sahara called the Sahel.[2]
The word *sahil* in Arabic means a plain, a coast, or a border.
Used geographically, the term refers to a band of territory approximately 200–400 kilometers wide, centered on latitude 15 N,
lying just south of the Sahara and stretching across most of the
width of Africa. The Sahel covers well over 2 million square
kilometers and constitutes significant portions of Senegal, Gambia, Mauritania, Mali, Burkina Faso, Niger, Chad, and even
the Sudan. By some definitions, the Sahel even extends into
parts of the Ivory Coast, Ghana, Benin, Togo, Nigeria, Cam-
eroon, and Ethiopia. The mean annual rainfall varies from about
100 mm in the north to about 500 mm in the south, and the
rainfall regime is highly variable. The rainstorms are erratic
and occasionally violent, and their variability increases from south
to north. The rainy season, lasting 3 to 5 months, alternates
with an extended, unrelieved, dry season. Drought is an inherent
feature of this harsh climate, and successive years of drought
may be followed by years of torrential rains. The region's climate
must be regarded as a variable rather than a constant.[3]

The soils of the Sahel are generally of low fertility, particularly
poor in phosphates and nitrogen, structurally unstable, with
low humus content and low water-retention capacity. Hardened
layers, laterization, and vulnerability to wind and water erosion
are common features. Water for irrigation is available in some

places from streams and rivers (Senegal, Niger, and Chari-Logone), and possibly from groundwater aquifers, but the area under irrigation is rather small and the irrigation potential has not been developed. The vegetation is a mixed stand of trees, shrubs, and perennial and annual grasses, typical of savannas and steppes.

A typical feature of arid regions in general is that the mode (the most probable) amount of annual rainfall is generally less than the mean: there tend to be more years with below-average rainfall than years in which the rainfall is above average, simply because a few unusually rainy years can skew the statistical average well above realistic expectations for rainfall in, say, seven out of ten years. More than 90 percent of the total variation in annual rainfall can be encompassed within a range between one-half and twice the mean. The variability in biologically effective rainfall is yet more pronounced, as years with less rain are usually characterized by greater evaporative demand, so the moisture deficit is greater than indicated by the reduction of rainfall alone. Timing and distribution of rainfall also play a crucial role. Below-average rainfall, if well distributed, may produce adequate crop yields, whereas average or even above-average rainfall may fail to produce adequate yields if the rain occurs as just a few large storms with long dry periods between them.

Drought is a broad, somewhat subjective term, that desig-nates years in which cultivation becomes an unproductive activity, crops fail, and the productivity of pastures is significantly dimin-ished. Drought is a constant menace, a fact of life with which rural dwellers in arid regions must cope if they are to survive. The occurrence of drought is a certainty, sooner or later; only its timing is ever in doubt. And it is during a drought that resource degradation in the form of devegetation and soil erosion occurs at an accelerated pace. Any management system that ignores the certainty of drought, and fails to provide for it ahead of time, is sure to fail in the long run. That provision may take the form of grain or feed storage (as in the Biblical story of Joseph in Egypt), or of pasture tracts kept in reserve for grazing when the regular pasture is played out, or of the

The Sahel region seems to have undergone a general decline
of rainfall since the late 1960s. There have been several unusually
prolonged and severe droughts since then, in marked contrast
with the preceding 20 relatively wet years. At each drought,
people tend to remain on the land in the hope that the rains
might soon return, and while waiting (and praying) they do
what they can to save their herds of goats, sheep, cattle, and
camels. They keep grazing their animals on the withered grass,
and when the grass plays out they try to increase their animals'
intake of browse by vigorously lopping trees already weakened
by lack of soil moisture. They also go on collecting firewood
from the sparse shrubs and trees, many of which die as a result
of this abuse. All the while, they continue to cultivate the dry
soil in anticipation of the rain. When many months elapse without
rain, the weakened vegetation dies out, while the soil—desiccated, pulverized, and trampled—begins to blow away in the
wind. And when a freak rainstorm visits the area, it scours
and gullies the erodible topsoil. Finally, the people are left
with no choice but to abandon their traditional homes and villages
and migrate to the cities, where they might hope to obtain
some relief assistance or find employment.

The damage to vegetation and soil typically results from
the combined effects of the prolonged drought and the excessive
pressure by too many people and animals on a land too parched
to support them. The same land that had supported them and
their forebears for so many generations is thereby in effect turned
into a desert. The tragedy of the American Dust Bowl is thus
being reenacted in Africa on a vastly greater scale of geography
and human suffering.

The proper management of an arid region is now the subject
of a lively and fateful debate in international development
agencies.[4] A term much bandied about in that debate is a region's
"carrying capacity." It is a rather nebulous term, intended to
characterize the amount of biological matter an ecosystem—or
rather, an agro-ecosystem—can yield for consumption by animals

or humans without being degraded. More simply put, it is a measure of how many people and/or animals an area can support on a sustained basis. Obviously, the productive yield obtainable from an area—and hence the number of people deriving their livelihood from it, at whatever standard of life—depends on how the area is being used; it is a function of technology. Under the hunter-gatherer mode of subsistence, an area may be able to carry only, say, 1 person per square kilometer, whereas under shifting cultivation it may carry 10, and under intensive agriculture perhaps 100 or more. The more intensive forms of utilization also involve inputs of capital, energy, and materials, such as fertilizers and pesticides, that come from outside the region itself but enhance its productivity. It is therefore highly doubtful that any given region can be assigned an intrinsic and objectively quantifiable property called "carrying capacity."

The figures often presented and quoted regarding the carrying capacity of the Sahel (and of other arid regions) are not rigorous calculations based on hard data. Instead, they are estimates drawn from whatever partial information is available, as judged by the presumed experts making those estimates. Moreover, they are usually estimates of regional averages, which obscure significant local variations in environmental, social, economic, and technological factors. Further, the inherent instability and uncertainty of rainfall in an arid region, and the ever-present possibility of drought, are such that estimations based on average conditions are likely to be over-estimations, and therefore poor indicators of actual, sustainable, carrying capacities.

One extremely important factor seldom taken into account in calculating a region's carrying capacity is the possible availability of water resources and the potential for making efficient use of them. These may include surface waters obtainable from either perennial or intermittent streams, or from the direct collection of runoff (called water harvesting), as well as from groundwater aquifers. Altogether, then, the carrying capacity of a region depends on how it is managed, and all that can be said in a given situation is that the particular region is either well-managed or over-exploited under the present mode of utilization, which may or may not allow realization of the region's sustainable potential.

The historic pattern, repeated over and over throughout the generations, was of gradual population growth and intensification of habitat utilization during favorable periods until a disastrous drought struck, resulting in starvation or emigration. The diminished population that remained in the area and survived the drought would then begin the cycle anew as conditions improved once again. The cumulative experience of many generations gradually brought about the evolution of mechanisms for the self-regulation of resource use, and for accommodation to the contingency of drought.

In the African Sahel, and similarly in other regions, the establishment and consolidation of European colonial rule in the nineteenth century brought about fundamental changes that subsequently were to modify the relation of indigenous societies to their environment. After an initial period of warfare, the area was stabilized and security conditions improved. So did medical and veterinary facilities, including vaccination services. These interventions allowed human and animal populations to increase rapidly and excessively during favorable periods. At the same time, traditional patterns of land utilization and tenure, and of migration and transhumance, were disrupted by arbitrary boundaries and by imposed political and economic structures.

The fortuitous occurrence of favorable weather conditions during most of the twentieth century, and particularly during the abnormally wet period of 1950–65 following the attainment of independence by the region's states, masked the effects of the changes imposed earlier. Given good rains, freshly cleared lands produced good harvests even in areas that normally would have been considered ill-suited for cultivation. Instead of deliberately keeping areas under-populated and providing for eventual drought, the authorities in some cases encouraged farmers to move into marginally arable lands. Pastoral tribes were then pushed further into even more marginal grazing lands, where they were provided with water by means of mechanically powered tubewells. Inevitably, drought struck. As access to the wells was free to all, traditional control over management of pastures was eliminated. The overall result has been an increase in herd numbers, a decrease in pasture through more widespread cropping, and abandonment of traditional range management mechanisms.

The drought of 1968–73 highlighted the basic problems that had been too long ignored. Family and tribal structures and their autonomous traditional practices of resource management and land tenure had been broken down, so the local population was now unable to cope with the drought on its own. The plight of the Sahel was exacerbated by the drought's recurrence, in even more severe form, during the early 1980s. Consequently, entire sections of the region were almost totally denuded of inhabitants, as hundreds of thousands of people walked away from their villages to huddle in relief camps and overcrowded cities.

Some of the Sahel's problems have been compounded by ill-conceived development efforts. Some planners seem to have misunderstood the logic of traditional production systems, and have underestimated the difficulty of improving them as well. They also failed to foresee the potentially negative consequences of intended improvements brought in under the imprimatur of "technology transfer." In many cases, they seem to have neglected the fundamental significance of rainfall variability, the probability of drought, and the principle of risk avoidance. Some of the traditional production systems were based on probable outcomes and were therefore better able to contend with droughts (though, of course, no production system can cope with a severe drought prolonged over several successive years). Most modern project designs, by contrast, have been based on average outcomes. The result has been that projects tended to incorporate overly optimistic assumptions that sooner or later were bound to clash with reality.

Agricultural planners working in the Sahel have tended to assume that the region's regular farming and grazing practices are basically viable in the long term, despite the fluctuations in rainfall. Their confidence was based on the fact that from

1900 to 1960 there had been recurrent droughts, but more-or-less normal rainfall usually returned within 1 to 4 years. Since 1960, however, although a few years were fairly rainy, the overall-rainfall became sparser and more variable. To cite one dramatic example, rainfall in Dakar had ranged from 500 to 800 millimeters in the 1950s, comparable to the American Midwest, but had fallen below 200 in the 1970s, comparable to the arid American Southwest. The effect of the long-term decline was exacerbated by shorter-term severe droughts, with practically no effective rainfall, that beset a huge crescent-shaped area stretching across the Sahel and doubling back through Southern and Southeastern Africa, an area twice as large as the United States of America.

Although the available historical records are rather meager, they suggest that similar major droughts, lasting 12–15 years, evidently occurred in the 1680s, the mid-1700s, the 1820s and 1830s, and the 1910s. In the first half of the nineteenth century, the level of Lake Chad apparently declined for two or three decades, to about where it was during the drought of the mid-1980s. The geological record shows several similar falls of the lake level in the past 600 years. On the other hand, we know that the Sahel has also gone through much wetter periods in the ninth through the thirteenth centuries, and the sixteenth through the eighteenth centuries; also from 1870 to 1895, and during the 1950s. The area near Timbuktu, which now has only 100 mm of annual rainfall, was humid enough in the late nineteenth century for wheat to be grown.

While drought cycles cannot be precisely predicted, and the evidence is not yet conclusive that human modification of landscapes can reinforce wet or dry tendencies, the increasing aridity observed in the Sahel over the last two decades is most worrying, especially in view of the fact that the population of the region has been growing much faster than ever before.

The population of the western African regions of the Sahel and the regions lying south of it, called the Sahelo-Sudanian and Sudanian zones, was estimated at 31 million in 1980. Though the population density is still fairly low throughout, varying from fewer than 2 per square kilometer in Mauritania to nearly 60 in Gambia, it has been increasing steadily. In recent decades, population growth rates have been close to 3 percent per annum, and they are expected to continue at that rate through the turn of the century. On this basis, the area will have 54 million inhabitants by the year 2000 (75 percent more than in 1980, and almost three times as many as in 1961). The urban population, incidentally, has been swelling at rates exceeding 5 percent per year, in large part from the influx of people driven off the land because of the drought.

The continued destruction of the rural environment is certain to result in further uncontrolled urbanization. As the demand for food, other agricultural products, and firewood will continue to mount, it will inevitably generate even greater exploitation of the region's meager resources. The best hope is to curtail population growth and to raise rural productivity by intensifying the use and conservation of favorable lands, developing the irrigation potential, improving management of range lands, reforesting marginal lands, and raising the efficiency of household energy use so as to curtail the burning of firewood. Above all, adequate provision must be made for the possible occurrence of drought in the future.

Fortunately, the land itself exhibits a remarkable resiliency. It had suffered many droughts in the past, and when the rains subsequently returned, so eventually did much of the vegetation. In large measure, the recent damage was temporary, and the land can recover if it is left undisturbed for a sufficient time. Satellite studies of the Sahel along the fringe of the Sahara show a generally southward retreat of vegetation in the dry period of the early 1980s, followed by a seemingly miraculous return of the vegetation during the rainy mid-1980s. However, intensive over-exploitation of the land during a very prolonged drought can undoubtedly result in some damage—in the form of erosion and loss of perennial vegetation—that is essentially irreversible, at least on the time-scale of human life.

What has caused the unusually severe and extended droughts that have visited sub-Saharan Africa during the last two decades?

Were they an entirely natural occurrence or has human activity somehow played a role? Any climatic phenomenon that extends over an entire continent and which had occurred intermittently in the past cannot be due solely to human activity, but must be related to the global atmospheric circulation that controls the climate of continents.[5] The question that remains is whether human activity might have worsened the effects of whatever natural factors may have been at play.

Various hypotheses have been advanced to explain the African droughts. One is that they are due to a cooling of the land masses of the Northern Hemisphere by about 0.3° C between 1945 and the early 1970s, owing to an increase in atmospheric dust from air pollution and volcanic eruptions. The cooling was presumed to have changed the pattern of air mass movement. Evidence in support of this hypothesis is sketchy, and seems to be contradicted by the heavy rains in the Sahel during the 1950s when the Northern Hemisphere cooled, and a second Sahel drought during the early 1980s when the Northern Hemisphere experienced a warming. Another hypothesis links the Sahel droughts to changes in sea temperatures in the tropical Atlantic, the so-called "Intertropical Convergence Zone"—a great band of equatorial clouds whose northward migration brings monsoonal rain to the humid tropics as well as to the Sahel. These changes might tend to reduce the northward penetration of the monsoon into West Africa and cause a decline in rainfall in the Sahel. A change in surface temperatures in the tropical Pacific, called the "El Nino/Southern Oscillation," has similarly been linked with droughts in the early 1980s in Australasia, India, South America, and Southern Africa, though these droughts did not persist for more than one or two seasons. Africa, on the other hand, has had many dry years that are not correlated with sea-temperature changes, so it is the persistence of the Sahelian drought that sets it apart from droughts in other parts of the world.

Still another hypothesis is that droughts can be caused or worsened by large-scale changes in the land surface of Africa, and specifically by the deforestation and overall denudation of the land. A process may thus have started whereby the drought can become self-reinforcing. According to the theory of "biophysical feedback," devegetation resulting from the drought as well as from overcropping, overgrazing, and deforestation, along with the consequent increase of the dust content of the air, combine to enhance the area's reflectivity to incoming sunlight. That reflectivity, called "albedo," may rise from about 25 percent for a well-vegetated area to perhaps 35 percent for bare, bright, sandy soil. As a larger proportion of the incoming sunlight is reflected skyward rather than absorbed, the surface becomes cooler and so air has less tendency to rise and condense its moisture as rainfall.

An additional effect of denudation is to decrease the infiltration of rainfall, and increase surface runoff, thereby reducing the amount of soil moisture available for evapotranspiration. Crops and grasses, which have shallower roots than trees and in any case transpire less than the natural mixed vegetation of the savanna, transpire even less when deprived of moisture during a drought. These changes may have some effect on local precipitation, since in many continental areas rainfall is derived in significant part from water evaporated locally. So the entire process can feed on itself as lower rainfall leads in turn to more overgrazing, less growth of biomass, and further reduction in reevaporated rain owing to the decline in soil moisture. Thus the biophysical feedback hypothesis offers its own explanation as to why the drought in the Sahel has tended to persist for so long. There is still no conclusive evidence, however, that even large-scale changes in land surface conditions do actually affect large-scale climate.

The notion that human environmental impact could cause changes in regional climate has long been viewed with suspicion by most scientists, but those views may be changing. While many climatologists still doubt that desertification could have triggered the Sahelian drought, some are now open to the possibility (albeit still unproven) that desertification may have contributed to the decline in rainfall and may explain why the drought in 1984 was much more severe than the one in 1968, though sea

surface temperatures and other conditions were similar in both
years.

The world's climate is likely to be even more influenced in
the future by human activity than it has been in the past. More
and more scientists, and lately the public at large, have become
aware of the possibility of a warming trend that might overwhelm
the cooling effect presumed by some to be caused by an increase
in airborne dust. That warming trend, or greenhouse effect, is
attributed to the release into the atmosphere of several radiatively
active trace gases, which have the property of preventing the
escape, and thus trapping, a growing proportion of the heat
emitted by the earth's surface.[6]

The predicted rise in the earth's surface temperature is likely
to increase the frequency, duration, and severity of droughts in
some areas, though no one yet knows enough to define just
when, where, and to what degree it will affect specific regions.
In any case, national and international agencies can no longer
ignore the need to plan for the contingency of drought, in Africa
and other dry regions.

*There is all Africa and her prodigies
in us.*

THOMAS BROWNE, *Religio Medici*

24

THE PLIGHT OF
AFRICA

NEARLY ALL THE PROBLEMS we have described find their most severe manifestation in the continent of Africa.[1] Although most of the people of Africa live on the land, many are losing the ability to feed themselves. The necessity to import more and more food has demoralized African societies and made some of them dependent on the largesse of donors. It has already saddled many of the continent's economies with a heavy and debilitating burden of debt, making it increasingly difficult for them to achieve independent progress. Nor is the problem Africa's alone, for not only is this continent the birthplace and seedbed of all humanity and a place of enduring cultural and geographic fascination, but it is also a region of great, largely dormant, ecological and human potentialities that the world can ill afford to neglect or jeopardize.

In the two decades following the Second World War, per capita grain production in Africa increased gradually, peaking in 1967 at about 180 kilograms per annum, a subsistence level considered barely adequate to sustain human well-being. Since then, however, total food production in the continent has increased at only half the rate of the population growth, and has

actually declined in some countries. In 1983 and 1984, per
capita grain production fell to about 120 kilograms, and as
many as 140 million of Africa's 530 million people needed to
be fed with grain from abroad. No other region in the world
has experienced such a retrogression in per capita food production.
And the problem is worsening: by the year 2000 sub-Saharan
Africa may incur an annual deficit of some 20 million tons of
food staples.

A 1981 report from the International Institute of Tropical
Agriculture, centered in Ibadan, Nigeria, summarized the African
dilemma:

> In Africa, almost every problem is more acute than elsewhere.
> Topsoils are more fragile, and more subject to erosion and degrada-
> tion. Irrigation covers a smaller fraction of the cultivated area
> . . . leaving agriculture exposed to the vicissitudes of an irregular
> rainfall pattern. The infrastructure, both physical and institutional,
> is weaker. The shortage of trained people is more serious. The
> flight from the land is more precipitate. . . . In one respect,
> namely the failure to develop farming systems capable of high
> and sustained rates of production growth, the problems of Africa
> have reached the stage of crisis.

The plight of Africa is an ecological crisis attributable directly
to the mismanagement of the continent's environment. Complex
interacting factors have contributed to this, the most fundamental
being the fragile nature of the environment itself, particularly
the soil; and the severe, prolonged, and recurrent droughts that
have afflicted the continent in recent decades and caused massive
famine and dislocation of entire societies. Added to these are
numerous interacting social, economic, and political factors, including a runaway rate of population growth, civil strife, armed
conflict, weak institutions of government, the failure of international cooperation, low prices for agricultural produce, uncertain
land tenure, lack of trained personnel, and—too often—wrong
advice from foreign experts and international agencies. In many
cases, the technology introduced into Africa from other regions
has proven to be inappropriate, while traditional local technolo-

gies that were appropriate in the past have not been adapted to accord with changing realities.[2]

Many Americans and Europeans have been impressed and moved by recurrent pictorial reports of the terrible famines that have afflicted several African countries in recent years. Most reports attributed the famines to drought, so the impression conveyed was of a severe but temporary natural disaster. However, the famines are not simply caused by a lack of rainfall. They are the end result of a long deterioration of the African agro-ecosystem, which has made the environment and people of that continent more vulnerable to droughts. So the problem is worse than it seems: for every emaciated person seen in the refugee camps, there are thousands suffering in their own villages, out of range of television cameras. And when the drought ends, the refugee camps are disbanded, and the international reporters go elsewhere. Those villagers, however, still remain with the problem of survival in an impoverished environment.

In 1984 and 1985, I had occasion to witness the effects of the drought on the land and people of Ethiopia and the Sudan. I was sent by the World Bank to recommend ways to improve irrigation and to provide drinking water for drought-stricken villages. All over the vast plains of Darfur and Kordofan in western Sudan and the Red Sea province in eastern Sudan, lands formerly verdant with lush grass and shrubs and trees were parched beyond recognition. I could see not a single green blade of grass, not a living shrub, nor a leafy tree. The windswept land lay bare and forlorn, prey to the sere winds of the desert. Carcasses of camels and cattle were strewn over the land, half buried in the loose dust and attended only by the vultures swooping down from their perches on the naked trees, whose trunks were lapped and polished by the swirling sands.[3] In village after village I found only a few dazed and frail people, too old or too young to trek across the desiccated plains and hills toward the already-overcrowded cities along the Nile. Nearly all the open reservoirs (called *khafirs*) designed to collect runoff were dry, and so were many of the shallow hand-dug wells. And yet I knew that the land could recover in time, provided only that it is not overgrazed or over-tilled.

A major cause of Africa's plight is the rate of its population growth. Pre-industrial societies have generally been characterized by high birthrates and high mortality rates (especially of infants), the two being in approximate equilibrium. The introduction of public health measures, and of improved hygiene and cleaner water supplies, has in the first phase of change tended to disrupt that equilibrium by reducing death rates and increasing the average life expectancy without an immediate and commensurate reduction of birthrates. Consequently, most societies entering the modern industrial age have experienced a spurt of population growth. Sooner or later, however, these societies enter a second phase, in which rising living standards and changing life-styles tend to bring about a spontaneous reduction of birth rates, so that population growth decelerates and populations tend toward stability, albeit at levels considerably above those of the pre-industrial period. Africa is still in the midst of the first phase, as a result of which its population growth, spurred by traditional social and religious concepts and by the immediate economic interests of families, now stands at 3 percent per year. This staggering rate, which implies a twentyfold increase per century, is the fastest of any continent in history.

Some observers suppose that population expansion, environmental degradation, and climatic change are reinforcing one another in Africa, and leading to an ecological catastrophe. Others believe, as does the author, that such a catastrophe can be avoided. The land-climate system is resilient and can recover. There are still large and as yet not fully utilized water and land resources. And although population growth must of course be controlled, the present population of Africa is not in itself untenably dense. In view of the present dismal situation, such optimistic statements might seem naive. But among the numerous failures there are enough successes and positive examples to prove that the continent's resources of soil and water can in fact be husbanded effectively. Africa is endowed with a great diversity of ecological conditions, and within that diversity there are regions with soils and climatic conditions suitable for the production of food crops, industrial crops, pastures, and wood products.

A brief description of Africa's land resources is in order.

The continent's soil and rainfall patterns have been compared to the layers of a huge onion.[4] The core of the onion is the Zaire (Congo) basin. Wrapped around this core, to the north and south, are concentric layers like onion-skins. The Zaire basin itself, and a band stretching along the coast of the Gulf of Guinea (from Gabon through parts of Nigeria, the Ivory Coast, Liberia, and Sierra Leone to Guinea), is the tropical rain-forest region. Here, the abundant rainfall (1,500–2,500 mm per year) has leached out most of the soluble nutrients and accumulated iron and aluminum oxides in the subsoil. The latter can become impervious (lateritic) hardpans when desiccated. North and south of the onion's core are semihumid regions (annual rainfall 800–1,500 mm) with soils that are less leached and hence somewhat more fertile, but which also may contain iron-oxide hardpans that make cropping difficult. The next "skin," stretching across Africa from coast to coast in the Sahel belt, and in the south through Namibia, Botswana, western Zambia, Angola and south-eastern Zaire, is semiarid (400–800 mm rainfall) with relatively infertile sandy soils that are low in humus and do not retain much moisture. The outer skin, encompassing the Sahara in the north and the Kalahari in the south, consists of desert sands. Finally, at the far north and far south of the continent, in the Maghreb and the Cape respectively, is the thin skin of Mediterranean-climate soils.

Within the overall picture, there are pockets of soils with exceptional fertility. Notable among these are the Nile Valley in Egypt and the Sudan, the volcanic highlands and their associated valleys in Ethiopia and in Rwanda-Burundi, parts of southern Nigeria, the vicinity of Lake Chad, and the highlands of Kenya. In most locations, however, the combination of soil and climate makes agriculture rather difficult to manage. Outside the tropical rain-forest belt, there are long seasons with very little rain and hot dry winds alternating with seasons of intense rains that tend to erode the soil and leach its nutrients.

The most dramatic human alteration of the environment in Africa has been the destruction of the natural tree cover. The rain forests of West Africa are disappearing at the rate of 5 percent annually. The Ivory Coast, which once had 30 million hectares of tropical rain forests, now has only 4.5 million hectares. Even more drastic has been the denudation of semiarid regions such as Mauritania in the west and Ethiopia in the east. Some 40 percent of Ethiopia was under forest at the turn of the century, 20 percent was still covered in the early 1960s, but only 2–4 percent could be discerned from satellite photographs in 1984. In the countries of the Sahel, in the so-called "homelands" of South Africa, and in the communal lands of Zimbabwe, vast areas are now almost entirely devoid of trees.

Accelerated soil erosion is a continent-wide problem, and is particularly severe in regions north of the equator, where it has apparently affected more than 85 percent of the agricultural land. Whereas the fertile soils of many other regions were formed on recent volcanic, alluvial, or eolian deposits, the soils of Africa are derived mostly from old parent material that has been long-exposed and hence is highly weathered and leached. Thus, most African soils are characterized by low productivity from the outset. Loss of topsoil not only deprives these soils of their limited reserves of nutrients but also increases drought stress due to reduced soil depth and low moisture-storage capacity, and to restricted root development. In Africa, annual erosion losses of from 10 to 40 tons per hectare are common on cropped and grazed land, and rates exceeding 100 t/ha/yr (equivalent to about 8 millimeters of soil depth) have been noted.

Erosion is most serious on denuded landscapes that have been bereft of plant cover, because of livestock grazing by nomadic herders who have no tenure to the land, and therefore no responsibility or incentive for its protection. Livestock numbers in Africa increased from about 295 million animal units in 1950 to about 520 million in 1983, even while the amount of land available for grazing diminished because of the spread of cultivation. Such an increase in grazing pressure has impoverished the grasslands, especially in the Sahel region. Perennial grasses have tended to die out and have been replaced by annual grasses that provide diminished soil cover. The loss of such trees as the acacias in the Sahel has reduced the quantity and quality of forage in the dry season, when the protein-rich pods of these leguminous trees could feed livestock on otherwise barren rangeland. The damage can become most acute during periods of drought, when the carrying capacity of the shriveled grass is diminished and the dried soil is laid bare and pulverized by trampling. The soil is thereby made prone to erosion by the sweeping dry winds as well as by the rains, which return sooner or later.

The process of erosion extends much beyond the Sahel: At least equally inclined to severe water erosion are the highlands of Eastern and Southern Africa. An extreme example can be seen in the Ethiopian highlands, where some 70 percent of Ethiopia's 45 million people eke out an existence by growing rainfed sorghum and a millet known as *teff* (*Eragrostis abyssinica*). Two-thirds of all Ethiopian farmland is on steep slopes of 20 degrees or more. Erosion in the highlands is not a subtle or gradual process, as it is in the Sahel. Gaping gullies are gouged out of the hillsides, and the soil and boulders that come tumbling down form fan-shaped piles of rubble in the valleys below. The denuded highlands of Ethiopia may lose as much as a billion tons of topsoil each year. Staggering erosion losses also take place in Madagascar. In Zimbabwe, almost half of the land cleared for cultivation in the last 50 years has been subject to serious erosion.

A sad example of countrywide erosion can be seen in Swaziland, that small and poor country nestled between Mozambique and South Africa. Eighty percent of Swaziland's 800,000 people inhabit rural areas, eking out a living based on subsistence farming and small-scale cattle herding. Throughout the country, gully erosion of sloping lands has taken on alarming proportions over the last few decades. A few fissures were noticed as early as the 1930s, but today the hillsides are scoured by thousands of deep gullies, called *dongas*, some of which are 30 meters deep and 500 meters long. They result from extensive deforestation, the clearing of shrubs, overgrazing, and cultivation. The soils are derived from volcanic rocks and limestone and are very erodible. Because of the undulating and in places mountainous character of the terrain and the irreversible process of erosion, a large part of Swaziland has already become a wasteland. Above altitudes of 1,000 meters, the gullies generally start in natural pastures, along the paths trod by cattle. They also form along dirt roads that are poorly maintained and poorly drained. Runoff and erosion, besides reducing the amount of arable and natural grazing land, also contribute to the silting of shallow streams.

The soils of sub-Saharan Africa are so low in plant nutrient reserves that what little inherent fertility exists in the root zone is depleted rapidly by crops and by losses in runoff and seepage.[5] As the nutrients removed by crops are not replenished by plant residues, manures, or fertilizers, African agriculture is in effect a soil-mining process. Even with yields as low as 1 ton per hectare (one-tenth the yield obtainable from fertile soils), maize and sorghum crops remove annually as much as 30–40 kilograms per hectare of nitrogen, 2–10 kg/ha of phosphorus, 5–30 kg/ha of potassium, and 5–10 kg/ha each of calcium and magnesium. Intensive, continuous cropping rapidly depletes soil nutrient reserves and lowers crop yields. Data from Nigeria indicate that maize yields can decline from about 5 tons per hectare immediately after forest clearing to no more than 0.5 t/ha after only four years of cropping, apparently owing to the combined effects of nutrient depletion, reduced content of organic matter, reduced storage of moisture, and deterioration of soil structure.

In addition to nutrient deficiency, soils of the humid tropics often present problems of aluminum toxicity.[6] The leaching of the relatively soluble bases—potassium, calcium, and magnesium—from the soil induces a condition of acidity in which hydrogen and aluminum ions predominate. High levels of aluminum are toxic to such crops as maize, beans, and other legumes. The condition can be rectified by application of lime, but where no economical source of lime is available the choice of crops is limited to those which are especially acid-tolerant.

The introduction of modern mechanized tillage into Africa for the purpose of increasing farming efficiency has, in many cases, achieved the opposite. Machinery-induced soil compaction can drastically reduce infiltration rates and lead to increased runoff, accelerated erosion, and declines in yield. Thoroughly pulverized soil toplayers are vulnerable to crusting, which is the spontaneous formation of a thin seal at the soil surface under the slaking action of raindrops. Such sealing can subsequently cause much of the rainfall to be lost as runoff, so the amount of moisture stored in the soil is reduced. The rain-slaked surface then hardens as it dries and forms a strong, brittle crust, which inhibits germination and seedling emergence.[7]

New land brought into agricultural use in Africa over the past seven decades has been estimated at some 100 million hectares. As people appropriated more land, natural forests retreated before the plow, and as trees no longer reabsorbed nutrients from deep soil layers and recycled them, soil fertility began to decline. With declining yields of grain and legumes, farmers have had to turn to tubers (potatoes, cassava, or yams) as their main starch source, in an effort to maximize the calories, rather than the protein, that their fields produce. The nutritional quality of their food has thus diminished. Moreover, as populations grew so did the demand for wood. In response, rural people cut wood from the remnant forests to use and to sell. As the trees disappeared, the situation got even worse: families now had to rely on crop residues and animal dung for heating and cooking, thereby further depriving the soil of nutrients. As each incremental parcel of cultivated land was rendered unproductive and barren within a few years of the removal of its natural plant cover, still more land had to be cleared to provide for the growing population.

To end this vicious cycle of destruction, the first step is to control grazing and the burning of vegetation, as well as the cultivation of marginal semiarid and steep lands. The next step is to reestablish and maintain vegetative cover. Shrubs and trees can be grown in association with food crops in an integrated system called agro-forestry. Cultivation and grazing should be limited to stable soils, where productivity can be improved and sustained by appropriate methods of soil and water conservation and of fertility enhancement.

Soil restoration requires improving the soil structure and its receptivity to rainfall by vegetative rather than mechanical means; by minimizing tillage and soil compaction, and by growing soil-improving crops and green manures. Additional requirements are to correct soil acidity and toxicity and to ensure a balance among the essential plant nutrients, and—most important—to prevent erosion by wind and water. These measures can succeed only if they are carefully and expertly directed and if sufficient incentives are given to local farmers to adopt, adapt, and maintain such practices. No one can expect hard-pressed farmers, on the thin edge of survival, to invest labor and capital in some future and uncertain prospect, unless they are given material and professional assistance, and an opportunity to witness the benefit, and unless they are given ownership of the land they may thus—and only thus—be prepared to improve.

Numerous well-intentioned efforts to remedy soil degradation in Africa have come to naught, for various reasons.[8] In 1977, the government of Ethiopia launched a massive effort to construct soil conserving terraces and drains so as to improve areas made barren by drought and over-exploitation. The campaign mobilized subsistence farmers and international assistance to construct contour embankments and ditches and to plant trees on 150,000 hectares of erosion-ravaged slopes. But of 500-million tree seedlings distributed for planting, scarcely 15 percent survived, owing to the combination of drought, careless work done in haste to fulfill quotas, and a poor choice of tree species, done without sufficient prior trial.

A vexing question, often asked, is why the famous "Green Revolution," which worked wonders in helping Asian countries to boost grain yields and to overcome hunger during the last three decades, has apparently failed in Africa. The answer lies in the fundamentally differing agro-ecosystems and socio-economic conditions of the two regions. In Asia, the primary form of agriculture is intensive production of rice and wheat as monocultures, generally under conditions of irrigation or abundant moisture that enable specialized crops to respond strongly and dependably to chemical fertilizers. A single package of successful yield-raising technologies could thus be adapted for use throughout the region.

In contrast, irrigation plays only a minor role in Africa, especially in its sub-Saharan regions. There, agriculture typically is subsistence farming and grazing under conditions of water scarcity and soil infertility, and without direct access to commercial sources of fertilizer or marketing facilities. Most African agriculture is characterized by a highly heterogeneous array of farming systems based on the production of numerous differing crops (corn, wheat, sorghum, millet, barley, and rice among the cereals; cassava and yams among the tubers). Much of Africa is semiarid, a fact that limits the profitable use of expensive inputs such as fertilizers. Hence the dwarf varieties of rice and wheat, and the specific techniques developed to produce them efficiently in Asia, have so far found little application in Africa. Whereas in the rest of the world gains in production in recent decades have come by means of intensification and per-hectare yield increases, Africa's lagging gains have come mostly from plowing more land—a practice that obviously cannot be continued much longer.

Another troubling question is why international aid and development agencies have failed to understand the special problems posed by the African environment, and to provide effective advice and help in promoting the sustainable economic development of the continent's countries. Too much of the assistance has gone to aggrandizing the capital cities of Africa with showcase hotels, factories, and public buildings, while the countryside has been neglected and allowed to deteriorate. The luxury hotels, filled with well-paid foreign development experts and salesmen, are surrounded by shanties with menially employed or unemployed refugees from the impoverished villages, factories either idle or producing goods few can buy, universities granting degrees to graduates who can find no productive work, and bloated bureaucracies and military establishments sustained by taxes and aid programs. The foreign experts spend much time writing reports, many of which are left unread and cease gathering dust only when overlaid by newer reports that are destined to suffer the same fate.

But reports unread may be preferable to reports that are implemented as programs, many of which ignore the needs of agriculture. Even programs aimed specifically at agriculture have often resulted in disappointment. Economic assistance strategies that were dictated by orthodox financial criteria (discount rates and expectable returns on investment) and managed by bureau- crats have failed repeatedly to answer the requirements of African agriculture. In their stead, Africa needs unorthodox and innovative approaches to arrest the continent-wide ecological deterioration. Such programs should transcend narrow project-by-project economic considerations and be aimed at comprehensive, yet flexible, environmental and social (people-based rather than primarily capital-based) development.

Many of the foreign experts who have guided development projects in Africa tended, perhaps unwittingly, to bring with them a northern or temperate-region bias. Many of them came from regions with abundant rainfall or irrigation, with deep and fertile soils, and under economic conditions that favored the production of cash crops as monocultures. This bias, and its corollary failure to recognize the particular environmental and socioeconomic circumstances of Africa, may have done more to retard the agricultural development of the continent than the much publicized droughts. It was exemplified early by the disastrous "Groundnut Scheme" undertaken by the British in Tanganyika in the years following World War II. The best of Britain's agriculturalists tried—and failed spectacularly over a period of 10 years and at an exorbitant cost—to produce peanuts on 1,200,000 hectares of land which did not get enough rain for that crop, and which turned cement-hard in the dry season. The result was that three very large sections of bushland were cleared and cultivated and made prey to accelerated erosion. Some three decades later, incidentally, an even grander agricultural development scheme was attempted by American investors in tropical Brazil, and the result was a proportionately greater debacle.

Unfortunately, similar large-scale failures have been repeated again and again by international agencies and commercial firms elsewhere in Africa. During the 1970s, a vast rice cultivation program was launched alongside the Niger River in the Mopti region of Mali. By 1984, rice could be harvested in only 10 percent of the area under development, and some 70,000 to 80,000 peasants were left impoverished. In 1970, the Canadian government sponsored a Tanzanian effort to grow most of the wheat it needed. Using mechanized equipment purchased from Canada and run on expensive imported fuel, 24,000 hectares were plowed up on the Hanang Plains. The Barbaig pastoralists who had long lived on the land were thereby relegated to poorer ranges, which their cattle quickly overgrazed. The soil in this area has a high content of clay, so it is able to hold water but gets very sticky when wet and shrinks and cracks as it dries. Severe erosion soon cast doubt upon the sustainability of the entire project. And so, yet another group of developers learned that the soils of Africa cannot be beaten into submission by methods of mechanized and commercialized production that are in conflict with the ecology and the socioeconomy of Africa.

The local traditions of African agriculture had distilled the experience and survival techniques of many generations. They were based on polyculture, the mixed production of indigenous crops for subsistence. Since many of the products were not sold commercially in monitored official markets, their value to African societies and economies has generally been underestimated. These traditions, which minimized risk, were brushed aside in the rush to create a market economy and to produce monocultural cash crops for export.

The largesse of donor countries dumping free surplus grain on hungry African countries, in the belief that it is the generous and noble thing to do, can in fact produce the cruel result of discouraging local production and hence exacerbating, rather than alleviating, vulnerability to famine. African farmers cannot compete when their markets are overwhelmed by cheap grain from abroad. The problem is not that local production is less efficient economically, but that the price of imported grain from the industrialized countries (the United States, Canada, Australia) is manipulated at the source, based on the profligate use of fuel that is kept artificially cheap, on subsidies, and especially on a disregard of the environmental consequences of burning fuel and of eroding or salinizing the soils. So much for the objective character of the so-called market forces.

Africa still has great potential for increasing agricultural production. According to figures compiled by the World Bank, only some 32 percent of its potentially arable land is actually used—a small fraction compared to the Far East and the Near East, where some 70 to 80 percent of arable land is cultivated. However, the greater potential lies not in expansion of cultivation onto infertile and erosion-prone marginal land, but in the efficient intensification of production on the best and most stable land. Although many of the continent's soils are of limited fertility, they can be improved.

Although Africa as a whole has lagged in improving land productivity, a few African countries have scored exceptional gains. Tunisia and Zimbabwe have each more than doubled yields in the last 30 years, and Egypt—though starting from a much higher base—has nearly doubled its yields. On the other hand, grain yields per hectare have fallen in such countries as Nigeria, Mozambique, Zambia, the Sudan, and Tanzania. In these countries, the beneficial effects of introducing improved varieties, and inputs such as fertilizers and pesticides, were more than offset by the loss of soil productivity.

Kenya offers one example of a promising combination of conservation and farming practices that increases the land's vegetative cover and reduces the likelihood of severe erosion. The Kenyan Ministry of Agriculture, with assistance from Sweden, subsidized farmers to terrace sloping land by leaving unplowed contour strips between fields and tree plantations. The terraces raised yields by retaining water and soil nutrients, and produced fodder for cattle. As terracing involves much labor for excavation and perpetual maintenance, which is difficult to sustain, trials are being carried out in several locations in Africa, under the auspices of the World Bank, to establish contour strips of dense vegetation (vetiver grass) instead of soil embankments.

Excellent work has been carried out at the International Institute of Tropical Agriculture in Nigeria to develop appropriate methods of soil management for humid and semihumid conditions. No-till methods, by which the crop is planted directly into the undisturbed soil without the conventional tillage for seedbed preparation, were found not only to save energy but also to improve soil structure and fertility, to enhance infiltration and storage of water in the soil, and—the final benefit—to increase crop yields.

Simple techniques of water-harvesting, of the sort developed in Israel's Negev, have been applied successfully in Burkina Faso since the early 1980s. For example, trees can be planted in a shallow basin scooped out of the soil, a few centimeters deep and a few meters across. During rainstorms, surface runoff water from a micro-catchment area, consisting of some scores of square meters, can be directed by means of low bunds to fill the basin and thus store water in the soil under and around the tree or trees so as to provide for subsequent needs. In this manner, fuelwood trees, shelterbelts, or even fruit trees can be grown in semiarid or even arid regions. Similar water-harvesting techniques can be used to improve pastures and to fill cisterns with drinking water for humans and livestock.

In principle, Africa has the capacity to feed itself and to attain a decent standard of living for its population. To achieve this, all the countries of the continent need to muster the will and the means to curtail population growth and to conserve their land, water, flora, and fauna. Although some land in Africa has undoubtedly been degraded severely, most of the arable land can be restored and utilized effectively. The African nations themselves can develop the leadership and organization to implement that task on their own behalf, and the international community should respond with adequate development assistance when called upon. Food aid can be a vital means of alleviating temporary famine, but it should not be allowed to create a condition of permanent dependency, nor should it inhibit the self-generated and sustainable development of the continent's own productive capability, coupled with protection of the environment. However bleak the situation may seem at present, it is a product of human mismanagement and should yield to human remedy, provided that remedy is well-conceived and implemented, and timely. The land and the people of Africa are resilient enough to overcome the present adversity.

*Can the rush rise without mire? Can
the reed grow without water?*

JOB 8:11

25

ENDANGERED
WETLANDS

THE TERM "WETLANDS" describes areas that are saturated, or
periodically inundated by a shallow layer of water, long
enough to have given rise to a characteristic aquatic community of plants and animals. As such, wetlands are distinct terrestrial ecosystems, often designated as swamps, marshes, bogs,
mires, moors, morasses, fens, and floodplains. In the stereotypical
view, these terms imply dampness, desolation, disease, and danger. The expressions "morass," "bogged down," "mired,"
"swamped," and "flooded" are often used to describe situations
that hinder, engulf, or overwhelm. Contrary to that prevalent
image, however, wetlands are not wastelands; they are in fact
among the most productive—and most threatened—of the earth's
ecosystems.[1]

Many wetland areas are being menaced by drainage, reclamation, development (agricultural, industrial, and urban), peat
mining, and pollution. Many have already been lost.[2] It is sadly
ironic that in attempting to exploit the riches of these habitats,
human societies have tended to destroy them. On the other
hand, we can draw new hope from the fact that more people
are beginning to appreciate the value of wetlands and the long-

215

term loss caused by the hasty destruction of these habitats for short-term profit.

Wetlands cover some 6 percent of the world's land surface and are found in all climates and continents, from the Arctic tundra to the tropics. Although it may be convenient to speak of wetlands as an entity, they are exceedingly diverse. Wherever they occur, however, they fulfill an essential role in the interplay of land and water, in the terrestrial cycles of water and nutrients, and in the disposition of organic and inorganic sediments. They sift, deposit, and decompose dissolved and suspended matter from floodwaters, transforming detritus into nutrients and thereby promoting growth. Many wetlands are more productive biologically than even the most intensively managed agricultural fields. Coastal wetlands, in particular, are sites of important physiochemical and biological interactions between saltwater and freshwater flows, supporting flora and fauna ranging from plankton to reeds, cypresses, and mangroves, and from shrimps to large and economically important fish. Wetlands serve naturally as spawning grounds for numerous species of fish and as breeding grounds for waterfowl and many other animals, and they contribute to the general ecological food chain even beyond their boundaries via currents and tides.

In addition to their role as natural habitats, wetlands support plants that are useful to humans, and offer many possibilities for aquaculture and for the combined production of fish and shellfish with timber and water-loving crops. Chief among these crops is rice—most of which is grown in flooded soils. It is the primary food resource of half the world's people and occupies about 11 percent of all arable land. Oil palm, a tree originating in West African wetlands, is one of the world's most important source of edible and soap-making oil, yielding more oil per year than any other vegetable and more oil than can be obtained from the fat of animals raised on the same area. Less well-known but also important is the sago palm, whose pith produces a starch that is the main food staple for a quarter of the population of Irian Jaya (the Indonesian section of the island of New Guinea) and over 100,000 Papuans. Wetlands are also a major source

of non-food plants, including: reeds and fronds of palms used
for thatch roofing, basket weaving, and paper production; hard-
woods used for timber, chipwood, and firewood; and extractable
resins of mangroves used for tanning. Many economically important animal products are obtained from wetlands: about two-
thirds of the fish and shellfish caught worldwide are hatched in
tidal wetlands. Especially important are such species as shrimp,
crabs, oysters, clams, prawns, and mussels. Wetlands produce
many of the world's valuable furs, such as coypu (nutria), muskrat,
raccoon, mink, and river otter.

Many species that depend on wetlands are being farmed,
an activity which may modify the fauna and flora of the wetlands
but does not require a fundamental change of the water regime
and of the environmental function of the habitat. Aquaculture
includes the production of such varied species as carp, catfish,
shrimp, crayfish, oysters, frogs, turtles, alligators, crocodiles,
and ducks. Average annual fish yields in fishponds are generally
of the order of a few hundred kilograms per hectare. However,
recent trials with fertilization and aeration of the waters have
produced annual yields of up to 5 tons per hectare, while results
from Israel suggest that harvests of 10 tons per hectare are possible.
Multiple cropping systems that combine fish and crayfish with
rice or timber can be especially productive.

In the developing countries, the urgent need to feed people
is often taken as a rationale for draining wetlands. But while
drainage may increase yields in the short run, it may also cut
down the ability of an area to produce its own sustained harvests.
Dams built to provide power and divert water for irrigation
often do so at the cost of downstream wetlands serving as wildlife
habitats or as fisheries. People too are affected. Lifestyles and
cultures have evolved around wetland cycles in Africa, for example, over many hundreds and perhaps thousands of years. The
seasonal migrations of people like the Nuer and Dinka of the
Sudan's vast Sudd swamp; of pastoral tribes such as the Tuareg,
Warbe, Sonhabe, and Peuhl of the inner delta of the Niger,
and the Ba Tswana of the Okavango in Botswana, are examples
of situations repeated throughout that continent's wetlands. The

environmental and economic cost of destroying such wetlands includes the loss of fisheries, of traditional grazing, and of culture. For many species of plants and animals, some not even classified as yet, destruction of wetlands spells extinction.

In the so-called developed countries, centuries of unbridled development have resulted in the irretrievable elimination of once extensive wetland habitats. Yet wetland conversion still goes on. In many cases, the natural regimes of rivers have been changed completely in order to achieve water control and to meet the ever-growing demands of expanding cities and industries. The seasonally inundated floodplain marshes, swamps, and forests of Europe and North America have contracted and been cut into smaller and smaller segments. They remain significant landscape elements in only a few exceptional areas, such as the Mississippi Valley and parts of the Danube basin, and as constricted fringes of smaller rivers that have somehow escaped 'total development' (meaning artificial regulation in the exclusive interest of promoting current human enterprises).

In the United States, more than 50 percent of the original wetland area has probably been lost since the Europeans arrived, and the remaining wetlands are being appropriated for use at a rate approaching 1 percent annually. From the 1950s to the 1970s, the average losses ran at about 185,000 hectares per year. Agricultural development has been by far the largest single cause of loss, accounting for 87 percent of recent wetland conversions in the contiguous United States. Urban development has taken 8 percent and other developments another 5 percent. Concern over these losses impelled President Carter in 1977 to issue an executive order aimed specifically at the protection of wetlands, and court cases since then have restricted the right of landowners to drain or dredge wetlands. Yet a complete national inventory of wetlands has not yet been completed and the principle of wetlands conservation is still too often ignored.

A prime example of a threatened wetland in the United States is the Everglades in Florida. Developers and politicians have tried to drain the vast marsh and turn it over to farmland ever since the annexation of Florida by the United States in 1821. Severe floods in 1928, 1947, and 1948 justified the con-

struction of 1,300 kilometers of levees and 800 kilometers of
canals. Some 16,000 hectares of wetland were lost directly by
the channel digging, and an additional 40,000 hectares were
drained. Flood control not only disrupted the natural flood cycle
and water regime of the Everglades, but also encouraged urban
and agricultural expansion. These changes led to declining populations of birds, fish, and crustaceans; to increases in estuarine
salinity; and to invasion by exotic plant species. The pressures
on what remains of the Everglades ecosystem, now largely contained within the national park boundaries, increased dramatically
in the 1970s, what with the uncertain water supplies now manipulated by the Corps of Engineers, expanding farmland, cities
demanding more water and reducing water quality, and the
threat of new development of adjacent lands. In 1983, the Florida
government launched a "Save our Everglades" project, and at
the present time it seems that this belated intervention has
actually helped to preserve a unique ecological resource.

Another example of development at the expense of wetlands
comes from Israel. Although the course of Israel's development
has, on the whole, been an admirable effort to balance the needs
of a growing economy with respect for the fragile nature of the
land and its scarce waters, the rapid pace of the country's economic
development has entailed some ecological mistakes. Arguably,
one of the most serious of those mistakes was the draining of
the Huleh lake and the associated papyrus swamp in the 1950s.
The motives for this project were to clear more land for agriculture, to reduce the mosquito breeding areas, and to save water
by reducing evaporation. The result was a regrettable environmental loss to the country. Only 310 hectares of the original 5,000-
hectare swamp were preserved. (Almost all 12,000 hectares of
the other swamps which existed in the country at the turn of
the twentieth century had been drained earlier.) Papyrus apparently reaches its northern limits in the Huleh swamp, and northern species like the white water lily are at their southern limit
there. Casualties of the drainage project include several plants
and over 30 percent of the original nesting-bird species, as drainage elsewhere left them no alternative sites.

The practical results of the Huleh drainage are also dubious, and perhaps more harmful than beneficial. The farmland added by the drainage has its problems: partly a chalky marl of low fertility, and partly peat which is difficult to farm owing to its tendency to oxidize, subside, and burn uncontrollably. The original expectations of saving water were only partially realized, as the vegetation growing on the drained land consumes almost as much water as the swamp itself did. Worse yet, the decomposition of the exposed peat has resulted in the release of large quantities of nitrates, which drain into the country's only remaining fresh-water lake, the Kinneret, and threaten its eutrophication. Finally, the Huleh—the entire lake and the swamp complex—might have served as an auxiliary perennial reservoir for water in case of need (such as a succession of drought seasons), so its hastily conceived draining might well be an economic, as well as ecological, loss to the country.

Many of the world's remaining wetlands are in Third World countries. They have survived thus far because they are in places that were difficult to develop, but are becoming increasingly attractive for agricultural expansion in poor and often famished countries. There is ever-growing pressure to convert to farmland the mires of Africa, South America, and Southeast Asia. In many developing countries the appropriation of wetlands is proceeding unhindered, and at an accelerating pace. In some cases the hunger for more agricultural land can lead to tragedy: the fertile new delta islands at the mouth of the Ganges draw Bangladeshi fishermen and landless farmers, some 10,000 of whom lost their lives on this precarious land in the cyclone of May 1985.

Other examples abound. The case of the inner delta of the Niger River in Africa epitomizes many of the worst problems facing wetlands. During the flood season, the inundated area supports more than 10,000 families engaged in fishing. During the dry season, the rich pasture of the floodplain supports more than a million head of cattle, and produces staple crops. Overgrazing, overfishing, and plans to build flood control structures may permanently limit the extent of the pastureland and fisheries. Planned upstream dams—as far inland as Guinea—could have a major impact not only on the inner delta but on the outer coastal delta, 2,500 kilometers away in Nigeria.

Apart from overuse and deprivation of water by drainage or upstream development, there is the further threat posed by pollution. In Pakistan, pollution and disturbance from agro-chemicals and industrial products pose a serious threat to the estuaries of the Indus delta. In Central America, pesticide contamination is a major problem in almost all the coastal wetlands bordering banana, oil palm, cotton, and rice plantations. The anthropogenic disturbances can greatly exacerbate natural changes, such as unfavorable climatic trends. The effects of the prolonged Sahelian drought on that region's wetlands is an extreme example.

A particularly large-scale transformation of wetlands will result from the planned diversion of the White Nile from the swamps of the southern Sudan. The purpose of the diversion is to save water now lost to evaporation, and to improve communications between northern and southern Sudan. The Sudd (meaning "barrier" in Arabic) is often described as the world's largest swamp. Within the Jonglei region of southern Sudan it consists of 11,000 square kilometers of seasonally inundated floodplain, sustained by flows from the White Nile and its tributaries. Remoteness and inaccessibility have so far protected the Sudd from development.

With its abundance of water and vegetation (including papyrus, Nile cabbage, and numerous aquatic grasses), the Sudd has flourished as a prime habitat for birds and large herbivores, and it is a particularly important link in the chain of sites along the Nile Valley for species migrating between the tropics and Eurasia. Included in its fauna are the endangered shoebill stork and possibly the largest number of water birds anywhere in Africa. There are nearly half a million tiang antelope, constituting one of the world's largest remaining populations of wild large mammals. Most of the wild populations of the Nile lechwe (an antelope of the genus *Kobus*) are also found in the Jonglei, as are buffalo, elephant, gazelle, hippopotami, white-eared kob, reedbuck, waterbuck, and zebra. The people of the Sudd and its environs—the Dinkas, Nuer, and Shilluk tribes, numbering 200,000 to 400,000—have developed life patterns involving pastoral migration, fishing, and agriculture, all closely geared to the cycle of seasonal flooding and ecological diversity of this unique habitat.

In 1978, construction began on the Jonglei Canal; at 350 kilometers it will be twice as long as the Suez. It is meant to provide an additional 4.75 billion cubic meters of water per year for irrigation and urban/industrial use, to be shared equally between Egypt and the Sudan. When finished, the canal will reduce the water flow into the swamp by up to 20 million cubic meters per day. It will then have far-reaching implications for the survival of the wildlife and the people who have long adjusted to the original wetland flooding cycles. Not only will the swamps and the floodplain shrink (perhaps by 50 percent), but the quality of the grazing lands will deteriorate greatly. The canal was originally scheduled to be completed in 1983, but because of technical and engineering difficulties, funding problems, and armed conflict in the area, construction has been delayed. By the end of 1984, only 260 kilometers had been excavated along a route from Bor to Malakal. The government of Sudan is determined to continue this project, but the ongoing rebellion in southern Sudan has so far made it very difficult to keep to any schedule.

An example of contrasting approaches to wetland utilization can be found in Indonesia. Massive infrastructure investments are now being made to reclaim coastal wetland tracts that planners formerly regarded as useless, but will now be used to absorb the transmigrants shifted from Java to the outer islands. Some Indonesian coasts, like the north coast of Java, have local population densities in excess of 3000 persons per square kilometer, and are among the most densely settled regions on earth. On the other hand, the coastal areas of Indonesia's outer islands, such as the tidal wetlands of southern Sumatra, have only 10–20 persons per square kilometer. There, wetland reclamation is resulting in large-scale environmental change. The same region has had traditional and successful small-scale wetland agricultural

systems that have existed for many generations in the very same
region. The traditional systems are polycultures of rice in association with coconuts—interplanted on mounds or ridges made of
soil that is removed from the shallow drainage ditches that separate
the rows of coconuts—together with mixed secondary crops of
great variety, including coffee, bananas, and vegetables.

Most government-sponsored schemes are based on a monoculture of irrigated rice. In their hurry to fulfill urgent food and
raw-material needs, planners have tended to ignore the traditional
systems, perhaps because of their impatience with small-scale
subsistence operations that they consider inefficient. Numerous
problems have beset the new scheme, including unstable water
supplies, saltwater intrusion, low soil-fertility, acidity, and aluminum toxicity; land subsidence resulting from soil shrinkage,
and the oxidation of peat (both inevitable consequences of the
drainage of swampland); pest infestation; periodic floods or
droughts, as well as difficulties in achieving social integration
and developing marketing facilities. These problems have made
it doubtful that the scheme can become ecologically or economically viable in the time the planners envisage.

Other developing countries have similarly ambitious schemes
that endanger wetland areas. Major canals are being built in
India to take water from the Brahmaputra, Ganges, and Indus
Rivers to drought-affected areas in Madhya Pradesh and Ra-
jasthan. Despite evidence of serious ecological and environmental
consequences, water-engineering schemes are being proposed that
could transform some of South America's principal wetlands,
such as the Rio Magdalena floodplain in Colombia and the once
remote Pantanal do Mato Grosso in Southwestern Brazil. The
latter is possibly South America's most important wetland. Covering more than 100,000 square kilometers—an area larger than
Hungary—it is a region of seasonally flooded savannas, with
scattered palms, many small freshwater lakes, and marshlands.

Significantly, China has some of the world's most extensive
and enduring systems of polyculture wetland utilization. These
systems, located in the Zhujiang (Pearl River) delta of Southern
China, are based on the husbandry of livestock, fowl, and freshwa-

ter fish (mainly carp), in combination with a range of perennial and seasonally rotated crops (including mulberry, sugar cane, litchi, longan, bananas, plantains, soybeans, mung-beans, peanuts, cabbage, sweet potato, maize, and many other vegetable crops). Pigs and ducks, together with fish, provide each household with animal protein and often a small cash income; while aquatic plants, crop residues, and kitchen leftovers feed the livestock, the manure of which (together with collected human waste) fertilize the fishponds and promote the growth of plankton—the principal food of carp. The ponds are 2–3 meters deep, and the crops are grown on dikes that are several meters wide and 0.5–0.7 meter above the water surface. The ponding depths and crop-bed heights, as well as the selection and sequence of crops, constitute an optimally controlled, diversified, and integrated system of production developed by trial and error over many generations.

In these systems, nutrients and energy are cycled continuously and little waste results. Several compatible and complementary species are raised in a synchronized sequence that permits the utilization of the system's productivity while maintaining a stable ecological balance. Once primarily a center of mulberry-tree and silk-worm production along with rice and sugar-cane growing, the region's production pattern has gradually become more elaborate, in response to both ecological and economic conditions. Because the Chinese system has sustained a large number of people for many generations, on a geographic and economic scale unmatched anywhere else, it can serve as a model of wetland management (to be modified as needed) for other countries as well.

Doth a fountain send forth at the same place sweet water and bitter?

JAMES 3:11

26

SWEET WATER AND BITTER

AS WATER IS SO VITAL to every ecosystem and every human endeavor, one might expect that modern societies would be mindful of the limited fresh water resources available to them, and take utmost care to ensure their continuing availability and high quality. Unfortunately, that is seldom the case. Because water seems so ubiquitous and readily accessible, it is regarded as a free good, abundant as air, endlessly renewable and therefore expendable without restraint. And because water is a fleeting fluid and much of it is hidden underground or readily vaporized into the atmosphere, it seems to have escaped the accounting books of many nations. The cost of drawing water from a source seems so low that few people are aware of the potential cost of replacing the water after the source has been exhausted, or of purifying it after it has been fouled. Much as they continue to degrade their soil resources, modern societies go on depleting and contaminating their water resources, in mindless disregard of the consequences.[1]

Those consequences can be witnessed in many places: once-perennial and plenteous rivers now dwindle at the end of the rainy season; once-brimming inland lakes are shrinking; wells

225

drilled into aquifers that were long thought to be inexhaustible are going dry. Worse yet, water supplies that were fresh and pure have been salinized or contaminated with toxic compounds that are gradually poisoning wildlife and threatening the health of humans.

The limited availability of fresh water already inhibits the economic development of many regions. Increasing competition for limited supplies has created tensions among regions and nations and may generate violent conflicts. The growing economic and environmental costs of providing water of adequate quality have called into question many of the convenient assumptions that have long guided our agricultural, industrial, and urban activities.

Great geographic and temporal disparities in water availability exist among regions. The shortage of water is, of course, most obvious in the arid regions of the world. We have already described the region of the Sahel, south of the Sahara. The African countries north of the Sahara are similarly drought-prone and short of water. The countries of Southwestern, Central, and Southern Asia suffer the same fate. Even in regions with an abundance of water, there are problems that constrain its use. In Africa, one of the world's great rivers, the Zaire—long known as the Congo, and second in volume only to the Amazon—accounts for about 30 percent of the entire continent's renewable supplies of water, but it flows through sparsely populated rain forest. In South America, the Amazon, which accounts for some 60 percent of the continent's runoff, is similarly remote from centers of population. In the Soviet Union, the largest rivers (Yenisei, Lena, and Ob) flow northward through the cold wilderness of Siberia to the Arctic Sea, and their diversion to the densely settled warm regions of Central Asia would entail enormous costs, both financial and environmental.

The inequities in potential water resources are exemplified by the extreme difference between Egypt, where the potential renewable water supply per capita per annum is less than 1,000 cubic meters, and Canada, where it is 100 times as much. Yet even water-rich countries may be subject to regional and temporal

shortages. In Canada, for instance, two-thirds of the river flow
is northward, whereas 80 percent of the people live along the
country's southern fringe. In the United States, there are decisive
differences in water availability between the Northeastern and
North-Central regions (including the Great Lakes region) on
the one hand and the arid Southwest on the other hand.

Historical records show that in 1900 the world's annual
water use was about 400 billion cubic meters, or 242 cubic
meters per person. By 1940, the total water use had doubled,
while the per capita use had grown by some 40 percent to
about 340 cubic meters. The rate of water use has accelerated
greatly since the middle of the twentieth century.[2] By 1970,
it had reached 700 cubic meters per capita. Both agricultural
and industrial water use grew twice as much in the 20 years
between 1950 and 1970 as they had during the first half of
the century. Today, the total annual withdrawal of water by
humans is near 4,000 billion cubic meters, equal to about one-
tenth of the total renewable supply of water, and about a quarter
of the stable supply available throughout the year.

Agriculture is the major user of water, accounting for about
70 percent of total withdrawals worldwide. Irrigated land now
amounts to some 17 percent of the world's cropland, and produces
over a third of the harvest. While contributing significantly to
the world's food supply, however, irrigated farming tends to
use an excessive amount of water, often more than twice as
much as the proven optimal rates of crop consumption, so that
a great latitude exists for water conservation in agriculture. Industry is the second major water-using activity. It includes water
use in power plants (both nuclear and fossil-fueled facilities),
as well as in food processing, mining, refining, and chemical
manufacturing. Unlike what happens in agriculture, however,
only a small fraction of the water used in industry is actually
consumed. Power plants, for instance, use water for cooling
and discharge it unaltered, except for the higher temperature,
into the stream or lake from which it was drawn. There is a
great potential for recycling water, as well as for improving
water-use efficiency in industrial processes.

The world's demand for water is likely to grow by at least 25 percent within the next 15 years. That is a rough estimate based on current population projections and rates of agricultural and industrial development, assuming that existing climatic conditions remain unchanged. Ironically, the population is growing fastest in some of the most water-short regions. The per capita supplies of water in Kenya and Nigeria, for example, are likely to diminish by close to one-half, and in India and Egypt by more than a fourth, by the middle of the first decade of the twenty-first century. Although industry now accounts for less than 10 percent of the total use of water in Third World countries, many of these countries are just beginning to industrialize, so their water requirements for power generation, mining, and manufacturing are certain to increase. In the foreseeable future, the demand for fresh water in quite a number of countries located in arid regions is likely to reach or exceed the total renewable or sustainable supplies, leaving no reserve for further economic growth, let alone for the maintenance of natural ecosystems. This situation underscores the importance of avoiding the wasteful practices now so rampant in most countries.

One illustration of the profligate way our society uses water can be found in our homes. In modern households with piped and pressurized water available at the turn of a tap, daily use is on the order of 100–300 liters per person. Affluent households with washing machines and dishwashers and lawn sprinklers often use more than 1,000 liters per day per person. Compare that with most of the developing countries, where water use seldom exceeds 50 liters per person! In arid regions, where women must walk for several kilometers each day to fetch water for their families in earthen jars or tin cans which they carry on their heads, water use per person is close to the biological minimum of 2–5 liters per day.

Modern agriculture and industry pose a dire threat to fresh water supplies beyond the problem of depletion. Residues of chemicals, including fertilizers and pesticides in agriculture and waste products of industry, have been finding their way into surface and underground bodies of water. Some of these chemicals

are toxic in trace concentrations, even when diluted down to
just a few parts per million, and in some cases even a few
parts per billion. Some of the industrialized countries have begun
to control the sources of these pollutants and to prevent and
even remedy the most acute cases of water contamination. In
most Third World countries, however, pollution controls are
either absent or inadequate to the task of preventing contamination by agricultural, industrial, and urban activities.

Normally, residues of natural organic compounds are decomposed by microorganisms in the soil. Waste materials can be
transformed thereby into nutrients or into harmless products.
However, in addition to these normal wastes, there are numerous
synthetic compounds that do not readily degrade under natural
conditions or in conventional waste treatment plants. Such wastes
or their derivatives can find their way into lakes and streams
from point sources such as sewers and drains, or from diffuse
sources such as fields and orchards.

Another threat to lakes is the enrichment of the water with
nutrients, a process called *eutrophication*. That process occurs when
large inputs of organic matter and mineral nutrients, particularly
phosphates and nitrates, lead to the accelerated growth of algae.
When the overabundant algae die, their microbial degradation
consumes most of the dissolved oxygen in the water, drastically
reducing the water's capacity to support aerobic life. Lakes that
were once clear can then become slimy and smelly marshes.
Experience in Europe and North America has shown that the
restoration of eutrophic lakes is possible, but the task can be
difficult and prolonged, involving the flushing out of excess
nutrients and restriction of their further inflows.

While the pollution of rivers and lakes is often reversible,
that of groundwater aquifers is much more insidious, for aquifers
are uniquely susceptible and, once fouled, may be practically
impossible to repurify. Groundwater is almost everywhere beneath
us, permeating the porous strata of the earth, an awesome natural
resource more vulnerable to long-term damage than the air we
breathe. With the slow patience that only nature knows, vast
quantities of groundwater wind their way sluggishly through a

labyrinth of infinitesimal fissures and pores, through cracked rocks and compacted layers of sand, sometimes taking a year to traverse a mere 10 meters, a human lifetime to travel a kilometer or two. A drop of today's rain that soaks through a permeable soil into an aquifer may not see daylight again in our lifetime or that of our great-great-grandchildren. When it does finally emerge to feed a stream or lake some distance away, it may carry the residues of the past into the far future. Hidden from view, groundwater can undergo pollution without anyone noticing. When finally the pollution arrives in a town's water supply, pumped from wells, a great hue and cry may arise from the aggrieved public. By that time, however, the damage will have long been done, and it may be too late to undo a process which had probably taken place over many decades.

Many organic pollutants enter groundwater as seepage from waste dumps, as leakage from sewers and fuel tanks, or as runoff from agricultural land or paved surfaces in urban and suburban areas. Because groundwater is cut off from the atmosphere's oxygen supply, its capacity for self-purification is very low. The microbes that normally break down organic pollutants need oxygen to perform their function. Pumping and disposing of contaminated groundwater is seldom practical. Prevention of contamination is the most rational approach, particularly where the groundwater reserves are shallow and hence especially vulnerable to contamination, and where they are used to supply potable water for domestic use.

Added to the industrial and domestic wastes are agricultural wastes. Important among these is the nitrate ion, a residue of fertilizers as well as a product of the microbial decay of organic matter within the soil. Because this ion is highly soluble and does not bind to soil particles, it is highly mobile and can readily migrate through the soil profile to the water table. Increasingly, nitrate has become a major pollutant of groundwater. In areas that are intensively cropped and fertilized, or have animal feeding lots and land disposal systems for sewage or garbage, the seepage of nitrates can cause the groundwater to reach concentrations in excess of standards established by the World Health Organization. Among other hazards, the excessive concentration of nitrates in drinking water may afflict infants with the "blue baby" syndrome—inadequate oxygenation of the blood known as *methemoglobinemia*.

The problem of nitrate contamination of groundwater[3] is epitomized in data from the Platte River Valley of Nebraska. There, the average concentration of nitrate nitrogen in the upper level of the aquifer rose from 3 milligrams per liter (equivalent to parts-per-million) in 1950 to 18 in 1980, an increase attributable to the growing use of mineral fertilizers to intensify the production of corn. An average of nearly 200 kilograms per hectare of nitrogen is added annually to most of the 17,000 hectares in the area concerned. The soils of this valley tend to be shallow and sandy, and the water table is now only a few meters, or tens of meters, below the surface, so the soluble fertilizer residues are readily conveyed to the aquifer by the excess of irrigation water that lixiviates the residues out of the root zone.

Even more serious contaminants than the residues of fertilizers are the residues or the derivatives of biological poisons known as pesticides. These include a bewildering array of insecticides, herbicides, rodenticides, nematocides, and so forth characterized by a high persistence within the environment. Their effects can reach much beyond the target pest organisms for which these materials were designed and applied. Many of them affect humans and may cause cancer, kidney or liver or lung disorders, and even birth defects.

The migration pathways of water and of contaminants from their point of origin through the soil and subsoil strata toward and into the aquifer, and their subsequent transport within the aquifer, are difficult to define and measure. The migration can be slow and spatially distributed, so that the contaminated water is detained in the soil and subsoil long enough to be at least partially decontaminated before it reaches the aquifer proper. On the other hand, the migration of contaminant-laden water to the aquifer can occur in rapid spurts through fissures or macro-pores that bypass and evade the soil's mechanisms of filtration or purification.[4]

The problem of aquifer contamination is compounded by the problem of aquifer depletion. Some aquifers hold water that is so old, and so slowly replenished if at all, that they are called fossil aquifers. If tapped, such aquifers may not be renewed in the foreseeable future. One example is the aquifer that underlies parts of the Sahara and the Near Eastern deserts and is the relict of an earlier humid climate. Even where recharge does occur, groundwater can be extracted at rates that exceed natural input, so water tables drop and water reserves are emptied out. Such overpumping, called mining, can support only a fragile and short-term prosperity, destined to end as the aquifer is exhausted, or as the water remaining in it becomes too expensive to pump or too salty or contaminated to use.

Such is the case of the Ogallala Aquifer in the United States. Over the last four decades, the amount of water withdrawn has exceeded 500 billion cubic meters, and hydrologists estimate that the aquifer is now more than half depleted under nearly one million hectares in Kansas, New Mexico, and Texas.[5] The fundamental question is why this important resource is being depleted at a time when the nation can afford to preserve it (at the very time, that is, that the government is paying farmers to idle rain-fed cropland in order to lessen a price-depressing surplus of crops). By using up the Ogallala Aquifer now, the United States is losing a reserve it may need in the future to meet vital food requirements domestically and abroad.[6]

Another example of aquifer depletion in the United States is the State of Arizona, whose current consumption of water from all sources is about 6 billion cubic meters per year, of which only about 3.4 billion are renewable. Arizona is mining its groundwater at almost twice the natural replenishment rate. The main culprit for this rate of depletion is irrigated agriculture, which uses nearly 90 percent of all the water in the state. For this reason, Arizona in 1980 became the first state in the United States to limit the pumping of groundwater. Undoubtedly, it will not be the last. Nor will it be able to maintain its present level of irrigation without massive federal subsidies for enormously expensive water transfer projects. However, the federal largess in constructing and subsidizing water supply schemes in the West, long taken for granted, is itself dwindling rapidly.[7]

Other aquifers in the United States and in numerous other countries are being similarly overdrawn. Notable among these are the aquifers underlying El Paso in Texas and Ciudad Juarez in Mexico, Dallas-Fort Worth in Texas, Tamil Nadu in Southern India, and China's northern provinces (including the environs of Beijing, the capital, and Tianjin, a major center of industry and commerce).

Aquifer depletion often produces a secondary effect—subsidence of the overlying land.[8] This phenomenon develops under two geological conditions. First, in carbonate rocks overlaid by unconsolidated deposits that receive buoyant support from the groundwater, the loss of that support can cause the capping layers to collapse suddenly into the emptied hollows within the rock, forming catastrophic sinkholes. Second, where saturated strata of clayey sediments overlie and confine a sand-gravel aquifer, pumpage from the aquifer is accompanied by vertical leakage from the clay layer, which then shrinks and causes land subsidence. Since the hydraulic conductivity of the clay is several orders of magnitude less than that of sand, the process of subsidence is a slow one, but as it continues over time it can do great damage to buildings, streets, and sewage systems, as well as to natural habitats. Land subsidence can range from as little as 0.2 to as much as 10 meters, and the area can extend from less than 1 to over 10,000 square kilometers.

Land subsidence has been taking place in such disparate locations as Beijing and Tianjin—where the surface has been sinking at the rate of 10 or more centimeters per year—Bangkok, Mexico City, and Houston-Galveston, where the land has subsided more than 2 meters in some 40 years. Most spectacular are the numerous sinkholes that have appeared in Alabama and Florida as a direct consequence of excessive pumping of groundwater and subsequent land subsidence. Thousands of these have occurred in the unconsolidated clayey deposits overlying carbonate rocks, mostly limestone. Their sizes vary greatly, from 1 to 90 meters in diameter, and from 0.3 to 30 meters in depth. The larger ones can easily swallow an entire building. Clusters of these sinkholes may create a landscape that resembles a battlefield, a stark illustration of human mismanagement of underground water resources.

Along coasts, land subsidence causes flooding, and a falling water table results in sea-water intrusion and salinization of the aquifer. Saltwater intrusion has already caused the contamination of drinking water supplies in a number of cities and towns along the U.S. Atlantic and Gulf coasts, as well as in many other countries.

Many of the major water supply projects around the world have involved the construction of large dams for the storage and diversion of rivers, for prevention of flooding, and for the generation of hydroelectric power. However, as many of the best sites for dam construction have already been appropriated, the rate of construction of additional dams has slowed. Moreover, some of the drawbacks to large dams have become increasingly apparent. Among these drawbacks are the submergence and consequent loss of riverine environments and of entire valleys with their forest, soil, and cultural resources; the displacement and cultural devastation of indigenous tribes; the rapid clogging of reservoirs by silt eroded from deforested watersheds, and deprivation of rivers downstream.

In recent decades, the potential to store surplus runoff underground has been receiving more attention. Artificial recharge of aquifers can be carried out by spreading water over land that allows it to percolate downward through the soil and subsoil, or by injecting it directly into the water table through wells. By these means, the underground reservoir is augmented, and the added water remains available for use when needed most, as during a dry season or a drought. Capturing excessive runoff during the wet season and using it to recharge the aquifer can convert damaging flood waters into a stable source of supply. Where the aquifer is contaminated with salts or other residues,

the introduction of fresh runoff can dilute the contaminants
and thus help to improve water quality. Underground storage
avoids some of the major disadvantages of surface storage, such
as submergence of valuable land, loss of water by evaporation,
silting, contamination by surface deposits, blocking of river flow,
and drastic disruption of the natural hydrological and ecological
regimes.

Some aquifers are recharged unintentionally by seepage from
unlined irrigation canals. Where this occurs, there is the possibility of using groundwater in conjunction with surface water to
augment water supplies, while preventing the water table from
rising to the surface and waterlogging and salinizing the land.
Such is the case in the Punjab province of Pakistan, which we
described in our chapter on irrigation.

One important country that is beset by an acute scarcity of
water is India, the world's second (and perhaps soon to be first)
most populous nation, with over 850 million people. Much of
the subcontinent is semiarid. Most of the rainfall occurs during
the monsoon season, typically June through September, so the
major task is to control flooding and to capture and store enough
water for the long dry season. Over the decades, the Indian
government has built numerous dams for the dual purpose of
storing water and producing electricity, but many of its efforts
are thwarted by the extensive denudation of the watersheds,
resulting in increased runoff, flooding, and silting, and in de-
creased groundwater recharge. Many streams that were perennial
now carry no water during the dry season. As a result, tens of
thousands of villages throughout the subcontinent have begun
to suffer shortages. Excessive pumping in the coastal districts
of Gujarat has caused saltwater intrusion into the aquifer, contaminating village drinking supplies.

India is engaged in disputes with neighboring countries over
water-related issues. Most of the water supplies of Bangladesh,
in the northeast, originate in India. Apart from the progressive
deforestation of India's northern highlands, which contributes
to the periodic flooding of lowland Bangladesh, there is India's
propensity to extract more water from the Ganges during the

low season, thus worsening the dry-season water shortage in downstream Bangladesh. Even more serious is India's dispute with its neighbor on the northwestern side, Pakistan. There the issue is the proper division of the water and the disposal of drainage in the arbitrarily divided Indus River Basin.

The problem of water is no less acute in China, the world's most populous nation. With 21 percent of the world's people but only 8 percent of its renewable freshwater, China's economic development is obviously constrained by limited supplies of water. The imperative to feed over 1,100 million people requires the high and dependable yields that only irrigated agriculture can produce. But growing cities and industries also require more and more water. Many of China's cities already suffer from an insufficiency of water. The situation is especially dire in the North China Plain, where Beijing and the important commercial center of Tianjin are located, and where millions of hectares of flat, fertile farmland demand to be irrigated. One possible solution is the northward diversion of the Yangtze River of central China, but this grand scheme is certain to be enormously costly and in any case will only postpone the looming crisis. Needed here— as elsewhere—is an all-out effort to boost the efficiency of water use by agriculture and industry, and to prevent pollution from reducing usable supplies.

Nowhere do the problems arising from the shortage of water, and the conflicts that shortage engenders, appear more acute and immediate than in the Near East.[9] Because of the dry climate, the region's countries depend largely on irrigation to provide food for their rapidly growing populations. The principal surface water sources in the region are the Nile, the Tigris-Euphrates, and the much smaller Jordan river. All these rivers are now in dispute, as conflicts over water are brewing among uneasy neighbors. Those conflicts are aggravated by the host of other contentious divisions—political, religious, cultural, and ethnic— already besetting this region.

Egypt, being a desert with practically no effective rainfall, depends entirely on irrigation to feed its 55 million people. Its sole source of water is the Nile. By agreement with the Sudan,

Egypt is entitled to a minimum of 55.5 billion cubic meters
per annum. However, other countries which control the Nile's
headwaters are not bound by any such obligations to either the
Sudan or Egypt. Ethiopia, for one, plans to use more of the
Blue Nile—accounting for some 80 percent of the united Nile's
total flow—within its own boundaries, so this tributary's contribution could diminish substantially in the future. That prospect
was recognized as early as 1970, when Egyptian President Anwar
Sadat threatened to declare war on Ethiopia over its intention
to build a dam on Lake Tana, the source of the Blue Nile.
Ethiopia still plans to dam the upper Blue Nile for irrigation
and electric power generation, but is constrained from doing
so because of its unending civil war and the lack of international
financing.

In any case, the Nile is subject to the variable pattern of
the region's climate. As related in the Biblical story of Joseph,
a periodic diminution of the river's discharge can visit great
hardship upon irrigation-dependent Egypt. During the drought
of the mid-1980s, the flow into Lake Nasser (the vast 500-
kilometer long reservoir formed by the Aswan High Dam in
the early 1960s) fell to 38 billion cubic meters per year, only
two-thirds of Egypt's allotted share. Since 1983, Egypt has had
to draw upon the reserves of that lake, so that by 1987 the
quantity of water in storage there had fallen to a meager 24
billion cubic meters—less than one-fifth the reservoir's capacity
and about a third of Egypt's normal annual water use. By early
summer 1988 the Nile levels fell to the lowest point in more
than a century. Tour boats that normally cruise the river's archaeological sites were left stranded on Cairo sandbanks. The drying
of the Nile threatened to break the slender fertile ribbon of
irrigated land that provides all of Egypt's home-grown food. If
the drought had lasted just one more season, Egypt would have
had to drastically curtail its generation of hydroelectric power
as well as its irrigation, and it—no less than the Sudan and
Ethiopia—would have required massive relief assistance to prevent
starvation.

At the last moment, in August 1988, Egypt was saved by

the sudden occurrence of heavy rainfall over the watershed far to the south. The results mixed redemption and horror in a capricious proportion. In Khartoum, the capital of Sudan, floods coming from the highlands of Ethiopia wrought devastation as the Nile exploded from its banks and inundated the city and its vicinity. And, as the river swelled and rushed northwards, it refilled Lake Nasser more than 16 meters above its level at the time of the drought in July, and up to the level it had in 1983. Egypt's water supply was restored, its generation of electricity could be maintained, and its crops could continue to grow under the blessing of irrigation. But that was a fortuitous circumstance that is not guaranteed to recur when another drought—perhaps even more prolonged and severe—befalls the region. The consequences next time could be catastrophic.

The Aswan High Dam has brought considerable benefits to Egypt's economy. For the first time, the annual Nile floods can be controlled. The dam forms a huge lake, which backs up some 320 kilometers in Egypt and another 160 kilometers in the Sudan. With an average width of some 20 kilometers and a maximum depth of 100 meters, the lake can hold an enormous volume of water. Its maximum capacity is 164 billion cubic meters, more than twice the normal annual discharge of the river. The water can be released as needed, to maximize its utility in generating electricity, promoting navigation, expanding the extent of irrigated land, and irrigating crops year-round. The lake itself affords possibilities for tourism and recreation, as well as for developing a fishing industry that may eventually compensate for the negative effects the dam has had on Egypt's coastal fishing along the Mediterranean.

The negative effects are more subtle. The first of these is the blocking of the fertile silt, which had long fed the plankton growing in the offshore waters along the Mediterranean coast beyond the river's outlets. Deprived of this rich plankton, the once-abundant sardines have diminished greatly. Moreover, the Nile itself, running clear of silt, has increased its erosivity and has been scouring its own banks. And along the estuaries of the delta, there is no more deposition, so the coast has been subject to progressive erosion by offshore currents. Consequently, the shoreline, which had advanced throughout history, is now gradually receding. A related effect of the damming of the Nile is the deprivation of Egypt's soils, whose legendary fertility had been renewed annually by the deposition of that nutrient-rich silt. To be sure, the blocking of the silt had begun long before the construction of the Aswan High Dam, as a consequence of earlier damming projects (including the delta barrages built in the nineteenth century and the Aswan dam built by the British in the beginning of the twentieth century). However, with the building of the High Dam, that blocking has been practically total. Ever since, Egyptian farmers have had to rely increasingly on expensive chemical fertilizers. At least equally deleterious is the effect of maintaining a nearly constant water level in the river, necessary to allow easy pumping of irrigation water throughout the year. This results in raising the water table and making drainage more difficult, so Egypt is now experiencing the maladies of waterlogging and salinization toward which it had for so long seemed immune. I have personally witnessed Egyptian farmers, beset by the high water table saturating their land, bring donkey-loads of sand from the desert to spread over their fields in a desperate effort to gain a few centimeters of height above the rising water, and thus stave off the waterlogging of their soils.

Overall, it is difficult to weigh the positive versus the negative effects of the Aswan High Dam, as the various effects pertain to different aspects of the Egyptian economy and environment and to different time periods. Hence they cannot be readily quantified. Suffice it to say that the benefits are immediate, whereas the drawbacks—not very noticeable at first—are likely to become increasingly important in the long run.

Even apart from the eventuality of a drought, Egypt is facing a dilemma that may soon become a crisis. Its population, now growing at the rate of one million every nine months, increases its demand for water and food from year to year. At the same time, some of its soils are suffering from the progressive waterlogging and salinization effects of the water-table rise resulting from the continuous maintenance of a high stage of the Nile. Yet doomsday scenarios are not inevitable. Much can be done to promote the efficient use of water and to augment water supplies. There is, however, resistance to changing traditional ways. The patient and faithful farmers have always believed that, thanks to God, the Nile has brought Egypt its precious water, and God willing it will somehow continue to do so. So they go on doing what they and their forebears have always done, working to divert water from the main canals to lesser channels that run like small veins between plots of vegetables and cotton.

Iraq and Syria are in a similar predicament. They, too, depend on exotic rivers: that is to say, rivers that derive their waters from another region. Both the Tigris and Euphrates originate in the highlands of eastern Anatolia (Turkey), from whence the Tigris flows directly to Iraq and the Euphrates courses through northeastern Syria on its way to Iraq. Both countries stand to lose part of their current supply when Turkey completes implementation of its grand Anatolia Development Plan. That plan anticipates construction of 13 hydroelectric and irrigation projects on the upper reaches of the Tigris and Euphrates rivers, including the massive Ataturk Dam on the Euphrates (slated to be the world's ninth largest dam). The dam may be a boon to impoverished eastern Turkey, but increased withdrawal of Euphrates water by Turkey will diminish the share remaining for the countries downstream, and could thwart their plans for further development of their own irrigated agriculture. Existing dams in Turkey already control the flow of water for Syria's Assad Dam, one of that country's primary sources of electrical power. Moreover, Syria's intent to draw more water from the Euphrates may deprive Iraq of half the water it has traditionally received from that river. If the three nations fail to conclude a water-sharing agreement by the time the Anatolia project becomes fully operational, Syria stands to lose as much as 40 percent, and Iraq up to 80 percent, of their potential supplies from the Euphrates. In January 1990, Turkey closed off the entire Euphrates for one month in order to begin filling the huge reservoir behind the Ataturk Dam. Turkey tried to compensate for this closure in advance, by increasing the downstream flow for 50 days in late 1989; however, Syria and Iraq will have to endure months more of flow restriction over the next five years before the reservoir is filled to capacity. The Ataturk Dam has thus emerged as the most visible symbol of the fateful competition for water that seems destined to dominate the international politics of the Near and Middle East.

Under the terms of two international conventions, the Helsinki Rules of 1966 and a U.N. convention of 1972, water rights are to be shared between nations according to population and need, with historical use patterns taken into account. But this formulation, though well intentioned, was left purposely vague, so it is difficult to adjudicate and practically impossible to enforce where serious conflicts of interest occur between uncooperative governments. Conflicts are especially difficult to resolve in view of the fact that international law also acknowledges the absolute sovereignty of nations over the resources they control. Thus, upstream countries such as Turkey can insist that they are under no more obligation to share "their" water than, say, Saudi Arabia is to share its oil. Hence the development of common water resources can lead to accord or discord: It can be either an opportunity to promote peaceful cooperation or a cause for conflict.

Syria's population, currently 13 million, is expected to swell to 22 million within 20 years. To meet the future needs of its people, Syria plans to irrigate an additional 40,000 hectares of land along the Euphrates, for which purpose it has constructed the Tabaqah Dam. The expected diminution of the river's inflow, along with the already worrisome depletion of Syria's groundwater reserves, may cause food shortages within the next decade. Even more tenuous is the situation of Iraq, which lies at the end of the Euphrates water line and will therefore be affected by Syria's, as well as Turkey's, water withdrawals. Particularly vulnerable is the wheat- and rice-producing region of northwestern Iraq. Unless mediated equitably and expeditiously, the water rivalry among these countries can indeed become a *casus belli*.

Jordan, Syria, and Israel are already in contention over water rights in the Jordan River basin. Syria and Jordan have coordinated plans to build a large Unity Dam on the Yarmouk River, a major tributary to the Jordan River. The project is intended to provide Syria with power, and Jordan with water to supplement that kingdom's insufficient irrigation system. What with the constantly shifting alliances of the Near East, relations between those two countries may or may not allow the plan to be implemented, but even if it is, a Syrian plan to divert water higher up the Yarmouk would give Damascus control over much of Jordan's irrigation-based agricultural economy.

To complicate the issue further, Israel is also concerned over the residual flow to the lower Yarmouk, from which it has for many years drawn water for its own vital needs. The Kingdom of Jordan itself projects a 50 percent increase in water needs within the next 20 years. A crisis could arise in that country before the end of the 1990s that would require transferring water from the agricultural sector to meet the expected domestic needs of a population now growing at a rate of 3.5 percent per year. Adding to the tension, Palestinians on the West Bank are vying with Israel for a larger share of common groundwater resources. Intensive well-drilling and pumping in the West Bank could reduce the amount of water available to Israel, which draws about a third of its supplies from an aquifer that originates in the country's highlands region and drains toward the Mediterranean Sea.

The most acute shortage of water is in the Gaza Strip, where population growth has overtaxed water supplies at an alarming rate. Underground aquifers in this region are capable of providing up to 60 million cubic meters of sustainable supplies annually, but the Gaza Strip's 650,000 inhabitants already pump some 150 million cubic meters, with the inevitable consequence of lowering the water table and inducing the intrusion of salty sea water into the region's wells. Continuation of the present rate of overpumping will surely result in an economic catastrophe.

Other troubled waters appear throughout the Near East,[10] notably in the states along the Persian Gulf, whose development boom has caused the rapid depletion of the Damman aquifer, their common subterranean reservoir. In the final analysis, the solution to the problems of the Near East, as of other regions around the globe, requires that the population explosion be curtailed and that an era of political and technical cooperation be instituted. When that happens, financial and human resources can be shifted from futile conflict to peaceful, environmentally sound, and sustainable development.

Turn our captivity, O Lord,
like streams in the dry land.

PSALMS 126:4

27

WATER MANAGEMENT
IN ISRAEL

ISRAEL OCCUPIES a sliver of land on the southeastern corner of the Mediterranean, perched precariously between sea and desert. Two-thirds of the country's area is in fact desert, and the remaining one-third is mostly semiarid. The climate is similar to that of California, which also suffers from a paucity of water. But there are important differences: Israel's population is about one-fifth, its area one-twentieth, and its water resources just one-fortieth those of California. Owing to its small size— only about 25,000 square kilometers—and extreme shortage of water, Israel has had to base its development on strict water conservation from the outset. For this reason, Israel has been forced to undertake the role of a pioneer in the management of scarce water resources, and the manner in which it has done so is of general interest.

The climate is characterized by two distinct seasons: a rainy winter, and a totally dry summer. While the general character of the climate is affected by the regional interplay between sea and desert, the topography of the land produces a heterogeneous pattern of local climatic differences that affect both the natural vegetation and the country's agriculture. The northern highlands,

244

especially their western (sea-facing) slopes, are relatively humid,
with an annual rainfall of about 800 mm. Aridity increases
from north to south, and from west to east. Accordingly, rainfall
varies north-to-south from about 700 mm down to 300 mm in
the coastal plain, from 800 to 400 in the central mountain
ranges (Galilee-Samaria-Judea), from 500 to 100 in the Jordan
Valley which is in the rain shadow of the central mountain
ranges, and finally, from 300 down to only about 25 mm in
the Negev Desert.

Israel's average annual water-supply potential has been estimated at about 1,600 million cubic meters.[1] In order to utilize
the surplus water of extra-wet years and to tide over the recurrent
drought years, the country requires a storage capacity of some
3,500 million m^3 for long-term regulation. Underground reservoirs—aquifers—are used for that purpose. At present, virtually
the entire potential water supply has been brought under control
and is utilized. About 70 percent of the water supply is used
for agriculture, especially during the summer months when most
crops depend almost entirely on irrigation for their water
requirements.[2]

Hydrogeologic exploration has been used for surface and
underground mapping of water-bearing strata. Isotopes have been
used to determine aquifer boundaries and recharge rates, and
computer-based models have served to simulate the dynamic
balance of water and to evaluate the potential consequences of
alternative water-use scenarios.

One of the first concerns facing the new State of Israel following its establishment in 1948 was the harnessing and allocation
of the country's limited water resources. Accordingly, a comprehensive legislative code, aimed at ensuring the efficient use and
equitable distribution of all water supplies, was prepared and
enacted in 1949. The first clause of that law established that
"the water resources in the State are public property, under
State control, and intended for the needs of its residents and
the development of the country's economy." With the right to
private or exclusive ownership of water thus abolished, a citizen
is only entitled to use water on condition that it be used efficiently

and in a manner that will not damage the environment or the rights of others.

A national Water Commissioner is empowered to manage the nation's water economy, with the prerogative to issue licenses for tapping or developing any source of water. Such licenses prescribe the quantities of water that can be extracted and the purposes for which they can be used. To facilitate supervision, the country is divided into rationing areas, in each of which norms of water use have been established for specific purposes. The country's network of thousands of wells is monitored continuously, to ensure that the aquifers are not overdrawn or polluted. Water charges are equalized throughout the country, and fines are levied for excessive or wasteful use. Water metering is enforced for all users, as a means of preventing wastage. Research-based guidance is provided on when and how much to irrigate. The entire water supply network is integrated so as to ensure elasticity of operation and dependability of basic supplies.

The northern part of Israel has 95 percent of the country's surface water resources but only 35 percent of the irrigable land, while the southern part has most of the arable land but is too dry to permit year-round cropping without irrigation. The Jordan River, which flows through the Sea of Galilee in the north, provides an average annual discharge of 480 million cubic meters. In 1956, a project was initiated to divert water from the Jordan system to the country's southern regions via a conduit known as the National Water Carrier. The undertaking was Israel's largest development project, involving the transfer of water over a distance of more than 250 km all the way to the northern Negev Desert, and it was completed in 1964. In summer, the water is used primarily for irrigation, whereas in winter it is used to replenish underground storage that becomes overdrawn in the dry season.

The country has two major aquifers: a deep limestone aquifer along the foothills of the central mountain range, and a shallow sandstone aquifer along the coast, extending over a length of about 140 km. The thickness of the latter is about 120 m near the sea, thinning out gradually toward the east near the

foothills. Both aquifers are replenished by rainfall. In addition,
the shallow coastal aquifer receives water through seepage from
the streams that cross the coastal plain along their path from
the mountains in the east to the seashore in the west. That
aquifer itself drains to the Mediterranean, discharging below
sea level. A useful addition to the country's overall water supply
potential comes from interception and damming of flash floods
which run off toward the sea following rainstorms in winter.
The captured water is spread over the coastal sand dunes, where
the infiltration recharges the underlying aquifer.

In addition to the conventional surface and underground
water resources, Israel has undertaken to reclaim and reuse municipal sewage for irrigation and industrial purposes. This involves
about 140 million cubic meters of wastewater annually. Where
brackish water is found, as in desert aquifers, attempts are made
to use it for producing specialized salt-tolerant crops.

Throughout the 1950s and 1960s, Israel's water planners
believed that the ultimate solution to the country's anticipated
shortage of water would be large-scale desalination of sea water.
So a great deal of research was devoted to developing efficient
methods for desalination, using techniques such as freezing,
flash-evaporation, electro-dialysis, and reverse osmosis. Pilot
plants were put into operation, especially in the Red Sea town
of Eilat. Plans were drawn to construct a dual-purpose nuclear
reactor, that would provide energy and also desalinate water.
The subsequent rise in the price of fossil fuel and of nuclear
reactors, along with problems of safety and pollution, have now
put a damper on the early plans for large-scale desalination.

In the meanwhile, Israel has perhaps done more than any
other country to develop and promote water conservation. Nearly
all irrigation water is delivered to the fields in closed conduits
and under pressure. Especially noteworthy are the methods developed to improve the efficiency of irrigation and crop-water use.
In most countries, irrigation efficiency is not much higher than
50 percent, and in many places it is even lower than 30 percent;
in Israel, however, irrigation efficiencies of 80 percent and higher
are commonly achieved. From the 1950s through the 1970s,

water applications per unit area have been reduced by about 25 percent, while crop yields have been increased dramatically and the value of agricultural production at constant prices has increased by more than 100 percent. Soil fertility and aeration are enhanced by specially developed techniques based on high-frequency, low-volume applications of water along with nutrients and pesticides. Research-based guidelines are provided to farmers on the optimal scheduling of irrigation for all major crops in the country's major regions. Time-controlled or flow-controlled valves and pressure regulators are used to maintain uniform delivery rates and calibrated quantities. Such systems can be automated and operated by remote control, using computer programs that adjust for variable temperature, wind, and crop growth phase.

Of particular interest is the Israeli development of drip irrigation,[3] whereby minute amounts of water, with injected nutrients, are trickled directly to the root zone of the crop to keep it continuously at a nearly optimal state of moisture, aeration, and fertility. The steady flow of water is obtained by using molded plastic drippers, inserted in weather-resistant plastic tubes that are laid on the ground or buried below the surface. Using this technique, irrigators can minimize losses of water in conveyance, evaporation, and deep percolation, and they can grow crops on soils formerly considered nonarable, such as sandy or even gravelly soils, sticky clays, steep and stony slopes, and even saline soils.

The need to expand its population and economy in the face of a strictly limited water supply has led Israel toward water-saving in industry and domestic consumption as well as in agriculture. During the period 1962 to 1974, industrial water use increased by 75 percent (from 55 to 95 million cubic meters), while the total value of industrial production rose by 220 percent. This increase in efficiency was promoted by the introduction of drier industrial processes, reclamation of industrial effluents, use of marginal water supplies and of alternative cooling processes, recirculation of water, and avoidance of careless wastage. The consumption of water in both private and public sectors has been influenced by education and publicity campaigns, and by

veyance losses by leakage. Finally, and very importantly, fiscal
measures—including the imposition of differential prices—are
used to discourage excessive water use.

Despite all of its efforts to date, Israel is facing increasing
problems of water shortages. With all its renewable water supplies
now overexploited, water is already being shifted from agriculture
to industrial and urban use. By some estimates, the country
may suffer an annual deficit as high as 30 percent by the end
of the 1990s, unless vigorous action is taken to achieve ever
greater levels of efficiency in water use.

In addition to problems of water quantity, there are acute
problems of water quality, particularly of groundwater pollution.
The problems include intrusion of sea-water and consequent
salinization of coastal wells; seepage into the aquifers of subsoil
irrigation percolates containing residues of salts, fertilizers, and
pesticides; and penetration of heavy metals, toxic organic chemicals, and microbial pathogens from a variety of sources, including
domestic septic tanks, sewage reservoirs, polluted streams, fish
ponds, garbage dumps, and industrial waste discharges.

The insidious problem of sea-water intrusion follows from
the excessive pumping of fresh water from the coastal aquifer.
Such pumping has caused the sloping interface between the
denser saline sea-water and the lighter sweet groundwater resting
on it to be shifted landwards by 500 to 3,000 meters. As a
result, many of the wells located near the shore have become
salinized and unusable as potable water sources. Equally pro-
blematic is the penetration into the shallow aquifer of fertilizer
residues such as nitrates, and of human and animal waste products.

Groundwater pollution, first noticed in the coastal aquifer,
is now detectable in the deeper limestone aquifer as well. Here,
rapid flow occurs in fissures and caverns dissolved in the limestone,
so there is little filtration and attenuation of the pollutants;
the groundwater is especially vulnerable to the penetration of
agricultural, urban, and industrial effluents.

Problems of water eutrophication resulting from increasing
concentrations of phosphates and nitrates have been noticed in

surface water reservoirs, by far the most important being the Sea of Galilee. Here, great and largely successful efforts have been made throughout the watershed to trap effluents and other wastes from settlements and fish ponds and prevent their flow into the lake. Still, the lake indicates a rising concentration of phosphates. While the problem is not yet acute, it is a cause for concern.

The story of water management in Israel illustrates how a nation cognizant of the problem can adapt to a situation of scarcity by improving efficiency. It also illustrates the complex and continuing nature of the problem, being one of water quality as well as quantity. Israel has innovated in developing water-efficient irrigation techniques and in creating an awareness and an incentive toward conservation. Still, the country cannot rest on its laurels, as its economy grows and as new problems arise, requiring an even greater effort to cope with them. Continuing and growing vigilance is the price of sustainable water management.

This even-handed justice
commends the ingredients
of our poisoned chalice
To our own lips.

WILLIAM SHAKESPEARE, *Macbeth*

28

ABUSING THE LIVING FILTER

O UR MODERN SOCIETY is addicted to manufactured chemicals. Among them are many that ease pain, alleviate disease, prolong life, spur food production, and serve as the catalysts for countless industrial processes. But once they are discarded into the environment as wastes, these concoctions, or their by-products, may be lethal. Our industries produce a prodigious mass of such chemical wastes, many of which are extremely harmful to many species and specifically hazardous to humans. Their safe disposal has become one of the most pressing environmental issues of our time.[1]

The traditional and time-honored practice of waste disposal is to apply it to the land. The premise underlying this practice is that the soil's evident capacities to retain, filter, decompose, and recycle the waste products of the natural biome can be utilized to rid us of our anthropogenic wastes. However, in far too many places, the amounts and kinds of waste materials currently applied to the land, including highly toxic chemicals, overwhelm the soil's limited capacities. The toxins can be buried, but not the problem. Instead of serving as the ultimate depository for the wastes, the overloaded soil has become a mere way station,

251

detaining the hazardous wastes instead of retaining or neutralizing them. Worse yet, in many places the soil has been so polluted as to have itself become a source of toxicity: and when eventually it releases the waste products added to it, it transmits them to surface waters and groundwaters, and thence into the food chain.

In the late 1960s and early 1970s, during the heyday of the new environmental movement in America, several commissions undertook to consider what problems might plague the nation's environment in the foreseeable future. They concentrated upon such important issues as air pollution (smog) and the eutrophication of surface waters, but most somehow failed to foresee the lurking threat of toxic wastes applied to the land. On the contrary, many touted the soil as the "living filter" able to absorb and cleanse all manner of noxious wastes (domestic and industrial, in their liquid and solid forms), and advocated increased application.

The shocking realization of that threat came only in 1976, in Love Canal, near Niagara Falls, New York. There, the Hooker Chemical Corporation had for some ten years (between 1942 and 1953) buried about 20,000 tons of toxic chemicals, many of which are known to cause cancer, birth defects, and a host of other maladies. When the dump was filled and no longer needed, the company covered it with soil and deeded it to the local board of education as a site for the construction of a school and of residences. More than twenty years passed before the problem came to light.

After several seasons of unusually heavy rains had raised the water table and inundated basements, houses began to reek of chemicals, trees and garden plants died, and children and pets began to exhibit chemical burns on their feet and hands. Then it was noticed that residents of the area had suffered an unusual number of manifestations of blood diseases, cancer, epilepsy, hyperactivity, spontaneous abortions, and congenital malformations. At least 82 compounds were identified in the soil, many of which were known carcinogens. Finally, in 1978, the site was pronounced unfit for human habitation. Several hundred families were evacuated, the school was closed, and the site was enclosed by a barbed wire fence. President Carter declared Love Canal to be a disaster area. In 1980, further testing revealed alarmingly high levels of genetic damage among people living in the vicinity, and an additional 710 families were evacuated.

The fate of Love Canal was only a first warning, and not an early one at that, of what has since been found to be a colossal national, indeed even a global, problem. Each year, American industry generates some 80 million tons of toxic waste, most of which is discharged onto the land and into bodies of water, and only about one-tenth of which is disposed of in a safe way. According to the Environmental Protection Agency (EPA), there are some 50,000 (and by other estimates, as many as 378,000) disposal sites in the United States containing hazardous wastes, and many of them are ticking time bombs that pose significant risks to human health and the environment at large. The practice of burying hazardous wastes in the ground has been going on for generations, on the seemingly plausible but untested and, as it turned out, wrong assumption that such wastes would lose toxicity during the decades it would take them to drift into aquifers through layers of soil and rock.

In contrast with wars and natural disasters—earthquakes or hurricanes—which strike swiftly and inflict their punishment in a manner that is unmistakable and spectacular, the unnatural disaster caused by soil-borne toxins is slow and, for a time, ambiguous.[2] Damaged chromosomes and brewing cancers remain unseen for years. The maladies strike at random, without a clear indication of cause and effect. We can surmise their activity and origin only by the statistics of chance, and only after they have been in effect for too long to prevent them from claiming numerous victims.

Thousands of potential Love Canals fester throughout the country, and these sites continue to infect ever-growing areas with their lethal toxins. Love Canal is unique only in that it happened to be discovered first. As the Love Canal situation became widely publicized, more and more dump sites of a similar nature have emerged. Times Beach, Missouri, intact with its streets and houses and a sign that reads WELCOME TO OUR

TOWN, welcomes the visitor to a ghost town, devoid of people. In the early 1970s, Times Beach had been polluted with dioxin, one of the most lethal chemical products, when a contractor spread contaminated oil on unpaved streets to keep down the summer dust. Subsequently, residents of the town began to develop various forms of cancer. It was not until late 1982 that they learned why.

Another town that experienced the chemical scourge is Holbrook, Massachusetts. The town park had long been used by a nearby pesticide factory as a depository for its chemical wastes. The area was strewn with half-buried and corroding metal drums containing residues of chemicals, including arsenic, DDT, and chlordane. The town was saved, though at great cost: the worst-contaminated ground was covered with a clay cap and the town's wells, all found to be contaminated, were closed.

Other examples abound. Each year, additional communities discover that they have been living near dumps or atop ground that had been contaminated by dioxin, vinyl chloride, PCB, and PBB,* as well as the more familiar but no less hazardous lead, mercury, and arsenic. Major sites have been discovered in Long Island, New York; Woburn, Massachusetts; Shepardsville, Kentucky; Luzerne County, Pennsylvania; Elizabeth, New Jersey; Baltimore, Maryland; St. Louis, Michigan; and Denver, Colorado. Smaller dumps where steel drums have been left to rust and leak, letting poisons seep into the subsoil for decades, are scattered in virtually every county of every state.

The political reaction to the problem was a "Superfund" created by Congress to clean up the thousands of leaking dumps that threaten to contaminate much of the nation's groundwater supplies. However, that fund has frittered away much of its resources in a piecemeal effort that only serves to reveal the enormity of the task.[4] Even sites that were purportedly cleaned up (generally by digging up and removing the soil) were only

---

* PCB, polychlorinated biphenyl; PBB, polybrominated biphenyl. These are only two among the bewildering myriad of organic pollutants, products, and byproducts of industrialization that are sometimes referred to as the toxic alphabet soup.

shifted to other locales, and these sites may in turn need to be cleaned up in the future. Thus, risks are transferred in a game of chemical leapfrog from one community to another or to future generations. And once the contaminants enter an aquifer, they become extremely difficult, if not impossible, to remove.

Organized land-disposal of wastes is usually carried out in landfills. The older ones were simply unlined and scarcely protected pits or valleys that eventually grew into artificial mounds towering high above the surrounding ground. In recent years, regulations have been promulgated requiring landfills to be underlaid, and sometimes overlaid, with clay liners. These are layers of clay, several meters thick, that are compacted for the purpose of containing the hazardous materials dumped into the landfill. Artificial liners of plastic or cement are sometimes added. Some modern landfills are equipped with drains and collection tanks, to prevent the noxious chemicals from leaking into the open environment. However, no system of containment is totally impervious, and many liners tend to develop cracks or yield to the corrosive action of potent chemicals. Some of the "leakproof" landfills installed today may become the hazardous waste sites of the next century.

Much hazardous waste is still improperly discarded, simply by dumping it wherever it is convenient or profitable to do so.[3] For too long, under the protection of its property rights, a company could discharge or bury its waste on its own ground with impunity, or pay someone else to do so on his land. Little consideration was given to the rights of future generations, and none at all to other affected flora and fauna. Nor was much consideration given to the tendency of hazardous substances to spread from their sites of deposition, via air or water, and to affect the larger open environment. Because of the slowness of natural biophysical and hydrological processes, these effects may not become apparent until many years later, with the attendant difficulty of obtaining a judgment against the responsible party. Meanwhile, unscrupulous operators have even taken to dumping poisons surreptitiously down sewers, into bodies of water, in forests or fields, and alongside roadways in isolated areas.

One possible remedy to these abuses is to enact regulations requiring that records be kept of the location of chemicals from production to disposal. The ultimate solution is to recycle rather than discard the byproducts, now considered wastes, and thus derive some benefit from them. The fractional amounts of these materials that cannot be recycled should, of course, be altered into harmless forms prior to being discharged. The know-how for such measures is already available in many cases, and the costs are likely to be less than those incurred by environmental damage. At present, however, society bears these costs, which should properly be charged to the perpetrators.

Chemical waste is only one particularly virulent aspect of the general land disposal issue in the United States. A problem of equal urgency is the disposal of garbage, which, though not as potent as toxic waste, is much greater in volume. Every year, Americans dispose of some 160 million tons of trash (up 80 percent since 1960 and still growing), not including sewage sludge and construction wastes. Each community is expected to care for its own trash, and most attempt to do so in landfills. Most if not all of these landfills leak their effluents, with numerous noxious materials carried in solution or suspension into groundwater or surface streams. But as the volume of refuse mounts, many landfills are filled to capacity. Many communities have been seeking to dispose of their waste by trucking it into rural areas, often across state lines, and some towns have begun to seek ways of exporting it out of the country. In the spring of 1987, the saga of the wandering barge laden with 3,000 tons of Long Island garbage vainly searching for a place to unload its fetid cargo, directed the attention of the world to a problem that had been festering for some time.

Municipal garbage contains many materials that can be recycled profitably. Among these are paper, glass, aluminum (and other metals), plastics, and readily degradable organics such as food and garden residues. The solution is to classify these components at the source and deliver them to recycling plants. The people of each community should be made conscious of the need, and given material incentives, to cooperate in this effort.

What cannot be recycled can be incinerated or composted for
subsequent use as fertilizer. Much can also be done to reduce
the volume of trash from the outset, by using less packaging
and fewer disposable goods.

Thus far we have considered the problem of land disposal
from the perspective of the United States of America, for that
is where the information is most readily available. But the problem
is worldwide. Some industrial countries have done somewhat
better than the United States. West European countries, with
their higher population densities and less available space, have
generally tended to rely less on land disposal and more on pretreat-
ment and incineration. The one exception is Britain, where hazardous and domestic wastes are lumped together (a practice called
codisposal) into landfills that are expected to be biochemical
reactors, able to degrade the noxious materials from both sources
simultaneously. While these landfills are monitored carefully
and are officially sanctioned, there is growing skepticism regarding their long term performance.

Some 200,000 to 300,000 tons of hazardous waste are shipped
annually from Western Europe to Eastern Europe, where it is
placed in landfills. This practice probably cannot be continued
for long, especially in view of the fact that Eastern European
countries are generating great quantities of their own hazardous
wastes and are beginning to suffer the consequences thereof. In
Poland, for example, high concentrations of heavy metals have
been found in vegetables, particularly in the heavily industrialized
region of Upper Silesia. Soil samples taken from vegetable farms
in that region have contained abnormally high levels of cadmium,
mercury, lead, and zinc. Inhabitants of Silesia reportedly experience significantly more cases of cancer and child retardation
than the population of Poland as a whole.

The problem of waste disposal is growing in the Third World
countries that are now industrializing and urbanizing. Although
the volume of waste per capita is still relatively low, few of
these countries have the advanced technology needed to dispose
of even this much waste adequately. The vicinities of such burgeoning urban centers as Cairo, Beijing, Mexico City, Calcutta,

Lagos, Bangkok, Sao Paulo, Jakarta, Karachi, Hong Kong, and Seoul, to mention just a few, are already experiencing some of the most ominous symptoms of contamination resulting from uncontrolled accumulations of wastes of all kinds. Attempts have been made to lure government officials in the less industrialized poorer nations to accept the hazardous wastes of industrialized nations, and such attempts have provoked justifiable resentment.

Japan, having been one of the first industrial countries to suffer the toxic effects of environmental contamination, now has the most advanced system for recycling industrial wastes. All levels of government and industry encourage waste reduction and recycling. There are facilities for the separate handling of oil, metals, tires, and plastics. Of the estimated 220 million tons of waste (including industrial and municipal) generated in 1983, more than half was recycled, some 31 percent was neutralized by incineration and other treatments, and only about 18 percent remained to be discarded in landfills. Still, the problem has not been solved in its entirety, even in Japan.

Efforts to reduce and neutralize noxious pollutants must be continued and intensified everywhere, lest more and more land and water be poisoned irreversibly.[4] Rudyard Kipling may have had a premonition of this future possibility when he wrote:

> For agony and spoil of nations beat to dust,
>   For poisoned air and tortured soil
>   and cold, commanded lust,
>   And every secret woe the shuddering water saw—
>   Willed and fulfilled by high and low—
>   let them relearn the Law.

UNTO

SOIL

SHALT

THOU

RETURN

UNTO

SOIL

SHALT

THOU

RETURN

Had we but world enough, and time

ANDREW MARVELL

29

A GLOBAL
ACCOUNTING

UMAN PRESSURE on the planet's limited and vulnerable
resources of soil and water began early and has intensified
steadily throughout the history of our species. Everywhere
we look, we see that these precious resources are now being
degraded, and in some places the degradation has already resulted
in crisis. At the same time that the people of the earth are
proliferating, their treatment of the earth is diminishing its
capacity to support them.

The environmental transformations we are witnessing are
driven by a continuous and accelerating increase in population
and land exploitation. More people exploit more land, so it
may then support more people, who must then exploit more
land, and so *ad destructionem*. Humanity has thus embarked upon
a self-generated, discordantly spinning double helix, with an
ascending spiral of consumption coupled with a descending spiral
of environmental exploitation, like a pair of whirling dervishes.
That frenzied *danse macabre* must now begin to wind down if
the ravaged biosphere with its diverse biota, of which humanity
is only a small component, is to survive more or less intact.

At the beginning of the eighteenth century, the world's

261

total population was about 600 million. Since then, it has increased by a factor of eight, with infant mortality reduced and average life expectancy prolonged. By the middle of the nineteenth century, the population had doubled to about 1.25 billion, and in the course of the next hundred years, it doubled again. It then doubled once more, to 5 billion, between 1950 and 1987. It is now expected to expand by another billion by the year 2,000. Notwithstanding the recent trend toward reduced fertility in some countries, the extraordinary momentum of population growth during the last few decades dictates that human numbers cannot start to decline in less than half a century. Even though the number of children per family might fall, the total population will continue to rise for quite a while, simply because of the increase in the number of young people of fertile age and in the overall life expectancy.[1]

Traditionally, most agrarian societies have had annual birth and death rates in balance at around 40 per thousand. In the last forty years, however, death rates have dropped in many developing countries to below 15 per thousand, while birth rates have remained close to traditional levels. The imbalance is most notable in Africa, where population growth rates in some countries are over 3 percent per annum. They were almost this high in Asia two decades ago, but most Asian countries are now undergoing a slow decline of birth rates that eventually will bring births and deaths into balance.

It seems obvious that until population growth is better matched to the resource base of each country, the sheer weight of numbers will generate mounting pressures on land and water resources. The dilemma is that population growth is both a cause and a consequence of poverty. Poor people often cannot avoid exploiting their environment. The pressing needs of the moment must be met from whatever resources they can muster, however limited or dwindling, even if the use of these resources jeopardizes their longer-term viability. The international community must help the poor nations to extricate themselves from this vicious cycle of population-poverty-resource degradation, so that a mutually reinforcing rise of prosperity levels and drop

of birth rates can work to turn a downward trend into an upward
one.

Although the growth of population doubtlessly contributes
to the environmental crisis, it is not in itself the major problem.
More important than human numbers is human behavior. Some
of the most densely populated countries in the world (such as
Holland and Japan) have managed their environments with greater
care, while others that are more sparsely populated (such as
the United States) have not done so well. Much depends on
each society's cognizance of the problem and its willingness to
regulate its use of environmental resources on a sustainable basis.
In this respect, America has been setting a rather bad example
of wastefulness and carelessness. One American, on average,
uses up more resources and produces more waste and pollution
than do ten Africans or Asians. It therefore ill-suits Americans
to preach restraint and responsibility to the world without setting
their own house in order.

In modern times, human economic activity has become increasingly global, with demands for goods and services in one
part of the world being met with supplies from half a world
away. Since the middle of the last century, nine million square
kilometers of the earth's surface have been converted into cropland. Energy use has risen by a factor of 80 over the same
period, with profound consequences for the planet's natural resources and chemical flows of carbon, sulfur, nitrogen, and other
elements.

The transformation of the planetary environment induced
by this explosion of human activity is particularly evident in
changes wrought upon the physical landscape. Since the beginning of the eighteenth century, the planet has lost six million
square kilometers of forest. Land degradation has increased significantly, as have the sediment loads of river systems. During the
same period, the amount of water withdrawn by humans from
the natural hydrological cycle has increased from about 100 to
some 3,600 cubic kilometers per year. In the last two centuries,
agricultural and industrial development have doubled the amount
of methane in the atmosphere and increased the concentration

of carbon dioxide by some 25 percent. Moreover, humans have begun to synthesize new chemical compounds and to insert them into the environment, and some of these have been found to exercise an unforeseen but powerful effect, even at low concentrations.

What were once local or temporary incidents of pollution or degradation now involve entire countries. The smog that once hovered over a few cities or industrial centers now envelops whole regions and acidifies rainfall over continents. Issues of economic growth versus ecological preservation that formerly were considered locally autonomous are now seen to involve multiple global linkages among energy consumption, agriculture, forest clearing, and climate change. Far from having made itself independent of the environment, humanity clearly is becoming ever more dependent on tenuous environmental relationships that at times seem too intricate to understand, let alone control.

It is worth noting that now as in the past a very high proportion of the world's population is concentrated upon a small part of its area. The major population concentrations—in East and South Asia, Eastern North America, and Europe—hold over two-thirds of the world's population upon only one-tenth of its land area. Compounding these differences in population density are differences in modes of life and impact on the environment. At one extreme, the richest 15 percent of the world's population consumes more than one-third of the planet's fertilizer and more than half of its energy. At the other extreme, perhaps one-quarter of the world's population lives in poverty and suffers periodic famine.[2]

The proportion of the earth's continental surface that is used for agricultural purposes may also seem to be small. In 1980, only 11.1 percent was used for growing crops; however, 23.8 percent was used for grazing livestock, including large areas of prairie, savanna and scrub vegetation. The picture looks different if we exclude the fraction of the earth's land surface that is desert, tundra, steeply mountainous, or otherwise unsuitable for human utilization. Excluding those areas, the fraction of the earth's available land surface currently utilized is more than 50 percent.

At the beginning of the twentieth century, the greater portion (some 75 percent) of the world's labor force was employed in agriculture, and the majority of these were subsistence farmers who produced mainly for their own needs. Before 1900, the main response to population growth had been to expand the total area under crops, while the intensification of production per unit area played a lesser role. Expansion involved clearing woodlands and advancing into areas that earlier were thought unsuitable for farming, such as uplands and marshes. During the same period, there were also major movements of population into new regions in China and Africa. The increase in cultivated land accelerated following the great European expansion and migration into the Americas, Australasia, Southern Africa, the grasslands of the Ukraine, and later east of the Urals.

In the twentieth century, however, the expansion of arable land has necessarily slowed, and the imperative of increasing yields by intensification of production has become the more important response to the rising demand for agricultural products. The twentieth century has also witnessed a growing interdependence between agriculture and industry. The rapid growth of cities resulted in an increased demand for agricultural products, food as well as industrial raw materials such as cotton and wool. On the other hand, farmers—once largely self-sufficient—now relied on industry for their tools, machines, fertilizers, pesticides, electricity, and fuel. As farm work became increasingly mechanized, the portion of the workforce employed on the farm declined greatly. In the European Economic Community as a whole it is now below 8 percent, and in such countries as Sweden, Switzerland, Britain, and the United States it has even fallen below 2 percent. Correspondingly, farms have become larger to exploit the enhanced labor productivity made possible by larger machines. Whereas on the eve of the industrial revolution it took four persons employed in farming to allow one to engage in non-agricultural pursuits, now one worker in American agriculture can sustain 50 or 60 in other jobs.

The vaunted productivity of modern agriculture has its dark side, however. The intensive use of fertilizers and pesticides (or biocides, as the latter ought to be called) has produced a growing chemical dependency, requiring greater and greater doses to compensate for the decline of soil productivity and the adaptive resistance of pests. Accumulating residues of these chemicals have had the effect of contaminating the larger environment, polluting groundwater and surface water, decimating wildlife, and threatening the health of domestic animals and humans. Larger machines, operated effortlessly and hence often used carelessly, can cause direct damage to the soil—including compaction and excessive pulverization, leading to accelerated erosion—as well as damage to aquifers by overpumping and depletion. Along with agrochemicals and mechanization, the introduction of high-yielding varieties made possible the specialized production of single crops year after year, a practice known as *monoculture*. This practice, involving the repeated pulverization of the soil in clean-till operations, has further exacerbated the problem of soil erosion. Nevertheless, monoculture has been promoted in the United States by government subsidies for such annual crops as wheat, corn, soybeans, and cotton.

In the decades following the Second World War, industrialized agriculture seemed to be phenomenally profitable, and was actively promoted by commercial suppliers of machinery and agrochemicals as well as by government and international agencies. Lately, however, the long-term profitability of this mode of agriculture has come into question, as its effects on the environment and on human health have become apparent. The enormous increase in labor efficiency resulting from the use of motorized machinery has been purchased at the cost of a greatly increased consumption of energy[3] and a reliance on external, unstable, and increasingly expensive energy sources. In some types of agricultural activity, the total amount of energy consumed—in fuel needed to operate engines and to produce and supply fertilizers and many other inputs—exceeds the energy value of the products. So while modern mechanized and chemicalized agriculture seems amazingly efficient from the point of view of human labor, it is remarkably inefficient from the point of view of energy. Judging by the latter criterion, modern agriculture is, in fact, less efficient than the mode of farming practiced in the unindustrialized coun-

sider primitive. Yet the growing cost of fossil fuels, and recognition of the negative environmental effects of their excessive use,
may induce a change in the mode of agriculture. Much of the
energy used in agriculture (for example, in excessive tillage and
fertilization) is in fact wasted. Far more attention must be devoted
henceforth to raising the efficiency of energy use in agriculture,
as well as to reducing the exaggerated application of potentially
harmful chemicals.

Consequently, efforts are being made to develop alternative
practices capable of enhancing the ecological stability and long-
term sustainability of farming by relying less on infusions of
fuel and chemicals from the outside and more on beneficial natural
processes and renewable resources drawn from the farm itself.[4]
Such efforts have been variously described as nonconventional,
organic, regenerative, or low-input agriculture. They are based
on diversifying and rotating crops (*polyculture* rather than monoculture), building up the soil by enriching it with organic matter,
promoting microbial activity that enhances the availability of
nutrients and the stability of soil structure, and using biological
in preference to chemical methods to control pests. Crop rotation
can reduce the incidence and severity of weeds, insects, and
diseases, as well as promote soil fertility—especially when it
includes the periodic planting of legumes that can help enrich
soil fertility by their symbiotic association with nitrogen-fixing
bacteria growing on their roots. Such legumes are most effective
in promoting soil fertility when they are incorporated into the
soil at the end of the growing season, a practice known as green
manuring.

Agriculture, being the employment of land and water for
human needs, is a fundamental aspect of our civilization, on
which rests not only the quality of human life but indeed its
very existence, in the future no less than in the past. If the
sustainable modes of soil and crop husbandry are further developed
and are adopted much more widely, agriculture may become
less intrusive and more ecologically harmonious. Until that happens, however, there is great danger that the opposite trend of

expanding mechanized and chemicalized agriculture will damage the larger environment and its natural biota, and ultimately destroy our civilization itself.

The global impact of human activity on the biosphere can be considered ecologically in terms of the fraction of net primary production (NPP) that humans have appropriated to themselves. NPP is defined as the amount of solar energy that is fixed biologically by photosynthetic plants (the primary producers), minus the amount of energy these plants consume in respiration. As such, NPP provides the basis for the maintenance and growth of all species of consumers and decomposers. It is thus the total food resource on earth. In principle, as humans take more of this resource, less remains for other species.

Recently an attempt has been made to calculate the fraction of NPP used by humans.[5] Three types of estimates can be used: (1) a low estimate based on the amount of NPP that people use directly, as food, fuel, fiber, or timber; (2) an intermediate estimate including all the productivity of lands devoted entirely to human activities, as well as the energy human activity consumes, such as in setting fires to clear land; (3) a high estimate including productive capacity lost as a result of converting open land to cities, and forests to pastures, or because of desertification.

The low estimate suggests that humans use about 5.5 percent of the earth's terrestrial net primary production. The intermediate estimate is 30 percent, and the high estimate is 39 percent. These figures suggest that people and the organisms that serve them already use a substantial fraction of the earth's productive potential, while the vast majority of other species are forced to make do with the remainder. The deprivation of other species and the progressive encroachment upon their realms result in the genetic impoverishment of the earth. The high estimate of the fractional NPP appropriated by humans suggests that humanity is one doubling away from usurping the greater part of the total biotic productivity of the planet. That is to say, if human numbers and economic activity double just once more (as they have doubled repeatedly over shorter and shorter time intervals in the last three centuries), little will remain for natural forms of life that are not directly of service to us. Humanity will then be living almost entirely in a world of its own making.

The above calculations, however, do not take account of the possible increase in the productivity of terrestrial and aquatic systems resulting from the best management practices already developed, and the potential for further improvements in the future. For example, the proven potential yields of agricultural crops are often 5–10 times the average agricultural yields now obtained in many countries. Production therefore can be greatly increased by improving the efficiency and ecological stability of current systems, without further massive expropriation of natural ecosystems. The intensification may include production of genetically improved plants in controlled environments such as greenhouse enclosures, with precise regulation of temperature, humidity, carbon dioxide, and nutrients. Such highly concentrated production systems are already known to be feasible, and when they are more widely adopted, their enhanced production per unit area may allow a considerable portion of marginal land, now being used improperly for agriculture, to revert to the natural ecosystems. However, controlled environment systems tend to be energy intensive, so their feasibility will depend on the attainable efficiency and future cost of energy use, as well as on the ecological stability of these systems.

On a planet that is 70 percent ocean, but with a human population already exceeding 5 billion and still growing at the rate of 1.7 percent per year, good land must be regarded as a limited resource. It must provide the foundation for human nutrition, the space needed for all other human activities, and— very importantly—the habitat and sustenance for all other terrestrial organisms. The fact that humankind, originally a child of the natural environment, has been able to gain mastery over the habitat, has in the final analysis made life in our environment more precarious rather than more secure.

There is an old adage concerning the difference between the clever and the wise: The clever are adept at extricating themselves from situations that the wise would have avoided from the outset. Unfortunately, our cleverness as a species has gotten us into a situation from which the same faculty can no longer free us. Cleverness has reached its limit. Wisdom is now needed.

A prevalent fallacy of governments and of international agencies is to assume that the needed wisdom resides primarily in the exalted profession of economics. Some economists, however, have dubbed their profession "the dismal science," while others doubt that it is a science at all. Outside the profession, a well-founded skepticism prevails. A common witticism has it that economists know the price of everything but the value of nothing. That is not a joke when it comes to ascribing a value to environmental resources. The term "economics" shares its origin with "ecology": both were derived from the Greek *oikos,* meaning house or home. "Economics" thus implies "housekeeping," or the management of the home environment. Yet our economists have contributed as much as any profession to the *mis*management of our earthly home.

Economists have typically sought to analyze investments in financial terms, using such concepts as discount rates and cost-benefit ratios, while relegating environmental factors to a secondary status, lumped together as externalities. Short-range calculations have granted license to one generation to profit from the depletion or degradation of natural resources, while ignoring the fact that it is bequeathing a diminished environment and a poorer life to its descendants. Apart from the profoundly immoral aspect of such a procedure, it begs a number of important practical questions. How much is an environmental resource really worth? Is it simply the cost involved in accessing and utilizing the resource, or should it include the cost of subsequently restoring it? How much do we stand to lose now if we forgo or postpone the use of a resource, and how much will we lose eventually if we do not preserve it? How much should we now invest in preserving it? And how might the benefit of its use, as well as the cost of its restoration, vary in the long run?

Since prices are often manipulated arbitrarily, and since the future can only be surmised from the past or from present trends which may not persist, it is practically impossible to make such economic decisions with any certainty. Nevertheless, some economists are now trying, however belatedly, to formulate an environmental economics adapted to unaccustomed long-term discount rates and to costs of resource replacement and prevention of environmental damage. A non-economist can only wish them success, while wondering if they still may not be missing an essential point about the environment. The point is that the value of a stable and viable environment exceeds its current economic worth or presently determinable price. Granted that economic considerations are germane to the management of the environment, they are only part of a larger set of considerations, including ethical ones, that are often neglected because they cannot be couched in financial terms.

The environment has an intrinsic value beyond all present-day economic measure. Just as we no longer trade with human lives or ask what slaves are worth, just as we have learned to accept the ethical principle that human life has inalienable rights and is not merely an economic object, so it must be with the environment which is, after all, the source and sustenance of life. The French statesman Clemenceau once said that war is too important to be left to the generals; we might say that management of the environment (for that is, in essence, what "economic development" generally implies) is too important to be left to the economists alone.

The failings of economists when acting by themselves in the environmental area are reflected in the activities of international development institutions such as the World Bank or the International Monetary Fund. These institutions exhibit all the symptoms of bureaucratization, tending to become self-enclosed worlds in which institutional growth and aggrandizement, and the internal interests of office holders, become ends in themselves. More disturbing is the defensive habit of rejecting inputs from scientists or other concerned groups, who are dismissed as academics, do-gooders, amateurs, or meddlesome activists. Such an attitude is viewed as arrogance by members of other professions concerned with the environment, and it leaves them with no recourse but to become vociferous critics. The resulting clamor of mutual recrimination is seldom constructive or enlightening.

The World Bank deserves special attention because of its leading role in international development. An unusually frank critique of its economic policy when applied to irrigation development, for instance, was given in 1988 by Dr. W. David Hopper, then senior vice president of the World Bank in charge of policy, planning, and research.[6] Prior to joining the Bank, Dr. Hopper had seen many irrigated lands become useless from waterlogging and salinization, and had therefore resolved, when accepting his responsibilities at the World Bank, that no irrigation investment should be approved without provision for drainage and a program of farmer education. Nevertheless, he was soon compelled to approve a project which lacked these essential components. The Bank economists had insisted that the project in question barely met the 10 percent economic rate of return, and that if drainage or education were added, the extra cost would make the project uneconomic. Moreover, the officials of the borrowing country were impatient to implement the showcase project as quickly as possible, unencumbered by environmental constraints.

Only a short time later, the newly appointed vice president had occasion to approve another project—adding a drainage system to irrigated land that had become waterlogged and saline after some years of prior irrigation without adequate drainage. This time the investment in drainage met the economic rate of return criterion because unproductive land was to be returned to productive use. But the costs were higher, much higher, than if the investment had been made at the outset. The anomaly of this story is that the same economic justification that permitted an investment that would surely degrade the land could later provide an acceptable rationale for making a much larger investment in the reclamation of land thus degraded.

The anomaly is even greater when one realizes that most irrigation projects still exclude investment in modern techniques of low-volume, high-frequency irrigation, in precise response to crop water requirements, as too expensive. Such techniques can minimize the problem at the source by *preventing* the over-irrigation that causes the water table to rise. Thus they can help to postpone the need for, and to reduce the eventual cost of, the drainage installation which is the subsequent cure for the self-afflicted malady of waterlogging.

What is basically at fault is the view that development should receive first priority, and environmental considerations are subordinate components of economic endeavors. This view is a perversion of the fundamental concept of development, which should mean improving the resource base rather than merely using the resource to generate a quick financial return on investment. Proper development should enhance the affected environment, or at least leave it unimpaired, rather than degrade it. In other words, it should be *sustainable* development. Unfortunately, this more fundamental meaning remains, in the words of Dr. Hopper, "a grail too often unnoticed by development practitioners."

The main function of the World Bank is to make loans (some $20 billion annually), for which it expects an appropriate rate of return. The Bank is a bank. By the Bank's Articles of Agreement, its staff are enjoined from allowing "non-economic" considerations to intrude on their analysis and appraisal of projects. Central to its lending policy is the concept of an internal rate of return (IRR), which each loan is expected to generate in excess of debt service costs. However, the IRR is frequently too low to cover the costs of environmental protection. Thus only the direct project costs are taken into consideration. The oft-repeated rationale for ignoring or delaying environmentally needed investments has been that their expected financial benefits cannot justify their additional costs. Moreover, officials of some borrowing nations may view environmental conditions as onerous impediments to accessing the resources of the World Bank and its sister financial institutions. Governments with limited tenure, in the developing as well as in the developed countries, generally respond to immediate political priorities; they tend to defer addressing the longer-term issues, preferring instead to provide subsidies, initiate studies, or make piecemeal modifications of policy.

Recently, the World Bank has attempted to rectify past

wrongs and embark on a new approach to the environment. Plans have been laid for conducting environmental assessments in various seriously affected countries, participating in the amelioration of pollution around the Mediterranean Sea, protecting tropical forests, dealing with desertification, controlling industrial pollution, reclaiming degraded land, and so forth. The good intentions are admirable, but concrete action has so far been lagging. For the most part, the Bank is still encased within its traditional economic *modus operandi*. It is still woefully deficient in environmental science and reluctant to enlist the participation of scientists in national and academic research institutions. This must change. The only plausible course for the future is to ensure that the environmental and developmental components are welded into a single set of considerations from project identification to implementation. Such an approach is preferable to requiring an environment impact statement to be prepared after a project is designed and ready for execution.

The problems of the time call for greater humility and mutual respect. Critics must acknowledge that even with all their shortcomings the international development institutions are motivated by sincere impulses and managed for the most part by well-intentioned and knowledgeable (though certainly not infallible) people. The policies and practices of these institutions arise out of a combination of human, professional, historical, and organizational factors that are difficult to change all at once. For better or worse, these are the institutions charged with the vital task of international development: we can never be sure that any other institutions in their stead would be more perfect. Environmentalists can scarcely afford to condemn or write them off; instead they should monitor the work of these institutions and take care to ensure that any necessary criticism be couched in constructive terms.

The wisdom we need will not be found ready-made in any single profession or organization. It can only be developed through interdisciplinary, interinstitutional, and international research and cooperation. And it must enlist the best and most concerned economists and social scientists, as well as environmental scientists and technologists. All of them must learn to communicate across arbitrary professional boundaries and jointly search for ways to provide tangible economic incentives for productive activities that also protect and improve the environment. As examples we might mention policies that penalize wasteful abuse of water and energy while subsidizing the adoption of techniques for more efficient use of these scarce resources.

The ultimate purpose of environmental activity should be to ensure that each generation transmits to its successors a world that has the range of natural wealth (enhanced, insofar as possible), and the richness of human possibilities that it received from its predecessors. Environmentally insensitive development can threaten that goal. We need to acquire greater awareness and knowledge, and more comprehensive understanding, of the interdependence between the environment and all forms of human activity, especially those that bear upon the land, its soils, its waters, and its varied forms of life.

I am not an optimist but a meliorist.

GEORGE ELIOT

30

A CASE FOR
CONDITIONAL
OPTIMISM

OVER THE LAST FEW DECADES, we have been buffeted by a
crescendo of crises that have shattered the once-prevalent
faith in an orderly civilized world and in the attainment
of a full life for all the earth's people. Among these are the
population explosion, pollution, biological extinctions, resource
depletion, famine, civil and international strife, ethnic and religious intolerance, poverty and drug addiction even in the most
prosperous societies, breakdown of social and personal mores,
new epidemics, terrorism and crime, arms races, proliferation
of nuclear weapons and nuclear accidents, energy shortages, and
the overall failure of international institutions to foresee and
forestall global crises and to close the widening gap between
the rich and the increasingly desperate poor nations. At times,
the difficulties seem so overwhelming that we instinctively recoil
from their very enormity. Is there hope for humanity on earth?
Can we find a way to live in harmony within a stable and healthy
environment?

Only a generation ago, most people would have answered
with a ringing affirmation of humanity's positive destiny. After
all, faith in the essential goodness of human beings and in the

276

efficacy of progress had been a fundamental tenet of our culture for many generations. Yes, many would acknowledge, there are serious problems, but they can be solved in time through education, research, technology, legal and social reform, planning, organized institutional as well as voluntary action, and the eventual attainment of international understanding and cooperation. For centuries, it seems, ever since the Renaissance and especially since the Industrial Revolution, the concept of progress had been the guiding principle of Western civilization. History was held to be a journey of progress, and every generation was expected to improve on the preceding one.

Now we are not so sure. It seems that our collective journey has taken us to some fateful boundary, beyond which lies uncharted and dangerous ground. Few of us can muster the equanimity to look on today's disaster-ridden scene as just a momentary setback in our voyage. The pessimists who sometimes seem to dominate the intellectual community predict the decline of our civilization, and expect us to be plunged into another dark age. They have transformed the prevailing perception of history from a bright promise to a threatening disaster. "We've seen the best of the game," declared the late novelist and critic C. P. Snow some thirty years ago. Even earlier, Arnold Toynbee had predicted that the developed countries would soon find themselves in a state of permanent siege, moral and psychological if not economic or military. Can rich first-class passengers on a train rushing toward doomsday go on eating and drinking while more and more third-class passengers clamber aboard and hang from the windowsills? Even if they disregard those now knocking on their walls, and barricade the doors so as to survive physically, can they survive morally?

Foremost among humanity's present plagues is the population-food-environment crisis. The world's population seems to be growing uncontrollably, especially in the underdeveloped nations of Africa, South America, and South Asia. Statistical extrapolations suggest that these nations could attain a population of 10 billion or more by the middle of the next century, unless famine, war, or self-regulation reduces these numbers. Meanwhile, food is already running short in some countries. And while the wealthy nations continue to feed grain to cattle, people in drought-stricken areas of the underdeveloped world cannot feed their children.

Increasingly, the answers searched for become regional rather than local, global rather than national. And there are no easy answers. Some may still believe that science and technology will solve all the problems in due course. But that naive and complacent assumption can only be called pathological optimism. Others believe there is no solution at all and that some nations are already doomed. That kind of pessimism is equally aberrant.

Especially dangerous are those who spread sophisticated reasons for despair. They say that the Green Revolution, which only 20 years ago seemed to be such a success, has run its course. They say that the world can no longer expand its agricultural production on a sustained basis; that major development schemes have failed; that the climate is changing imminently, and certainly for the worse; and that pollution, degradation, and depletion of physical and biological resources are inevitable consequences of economic development and of population growth, which itself is an ingrained tendency of humans that cannot be controlled except forcibly. Moreover, they say that food production will soon begin to decline, as soils will continue to degrade and as the worsening scarcity of energy will drive the cost of fertilizers, pumped water, pesticides, tillage, and mechanized transportation out of the reach of many marginal farmers. Therefore, they say, humanity is condemned to a life of lesser quality and security in an impoverished world of diminishing possibilities.

There is, however, an alternative view, which we might call conditional optimism. Though fully cognizant of the extremity of our situation, the conditional optimist draws a different picture of the possible future, yet one that is at least as plausible as that of the pessimist. Conditional optimism emphasizes that the rate of population growth has begun to slow down, because of government programs and changing precepts and lifestyles;

and that this trend can be enhanced by an active policy of education and of social and economic inducements. Further, while the earth's potential to support people is certainly not unlimited, it can suffice for a considerably larger population, provided the proper practices (the principles of which are already well known) are more widely adopted. The current shortages are due more to poor management than to any fundamental scarcity of resources. Present yields are extremely low in many of the developing countries, and as they can be boosted substantially and rapidly, there should be no need to reclaim new land and to encroach further upon natural habitats. Moreover, the proven possibilities for developing intensive, efficient, and sustainable agriculture and industry can obviate the need for the widespread cultivation and grazing of marginal lands, thus allowing the regeneration of natural habitats.

Environmental degradation is not dictated solely by population numbers, but also—and more importantly—by human behavior. Now that this principle is recognized, individuals and nations may begin to change their ways. Pollution is not irreversible; it can be controlled once the problem is recognized and the will and means are mustered to remedy it. And while we cannot expect the rich nations to give up their wealth or the poor nations to wrest it away by force, we can expect a greater measure of cooperative international action to alleviate hunger by improving food production and distribution, and to safeguard the common environment.

The history of humanity has always been a race between learning and disaster. The form of the ever-looming disaster has changed repeatedly, like a many-headed dragon. Long ago it was famine, then pestilence, then degradation of the environment; then once again famine, new outbreaks of disease, and still more degradation. None of these afflictions can ever be cured once and for all. Each new or recurrent problem requires a higher level of knowledge and more concerted action. Only by acknowledging and acting on the underlying causes can humans head off the threat of disaster. This belief was a common article of faith during the first half of the twentieth century, but is seldom heard today in respectable forums, though there is still much to be said for it.

Conditional optimism is not a license for complacency or an excuse for inactivity. On the contrary, it posits that much can and must be done, and that the time is ripe for action. Its optimism is conditioned on such action being started without delay. The necessary tasks will not be easy to sustain.

Perhaps what has gone wrong is not the principle of progress itself, but our current distorted perception of it. We now seem to define it as having less to do with quality than with quantity. We seem to have developed an insatiable and profligate appetite for more and more material possessions. We have become careless wasters of resources, compulsive seekers of comfort and of luxury. Our sense of pleasure is similarly distorted. Most of us no longer seek the pleasures of communion with the natural world but the artificial thrills of contrived entertainment. Our detachment from nature, itself a perversion, serves to further pervert our reality.

Yet subliminal ancestral memories of a more natural life call us back. We sense a vague yearning to return to a life of closer connection with the land, the trees, the waters, the sky and earth, and the seasons.

A new climate of opinion is forming. More people are becoming aware. There is spreading and deepening recognition that the growth rate of the human population must be curtailed. Many now realize that we must protect not only the physical environment but also its diverse community of species co-inhabiting the biosphere. We know we must reduce our wastes and find safe means for their disposal. We know we must not degrade or deplete the resources upon which our children and grandchildren will inevitably depend. Science can illuminate these issues and lay the basis for a technology to deal with them, but only our determined commitment to control ourselves can save us.

A major challenge of our time is finding ways to achieve harmony between the responsibilities and needs of developed nations and those of the underdeveloped world, between the needs of our generation and those of future generations, and between the human species as a whole and other species. Ecology teaches us that each member of a community is defined not by its individual characteristics alone but by the nature of its reciprocal relationships with other sharers of the same habitat. The ancient tribal vision of the world is still deeply ingrained in us. Gradually, however, our vision has evolved and our notion of kinship has been extended to include first our immediate clan or tribe, and then successively our village, city, country, race or creed, and—eventually—all of humanity. This expanded perception of kinship and allegiance must now transcend the human species and extend to the totality of life on planet Earth.

One hopeful sign of change has been the growth of an environmental movement, which has already played an important role in altering our values. The necessary changes in behavior must begin with individuals and small groups, and ripple outwards to encompass larger numbers. Public opinion can thus be awakened and mobilized in behalf of environmental causes on a global scale. Though the proliferation of organizations and the din of numerous conferences, each vying for attention, may seem excessive at times, their underlying purposes are serious and positive. Lately, even politicians have begun to express concern for the environment. The excellent work of the World Commission on Environment and Development (chaired by Gro Harlem Brundtland, prime minister of Norway), embodied in the publication *Our Common Future* (1987), is a significant milestone. Another landmark achievement is the 1987 Montreal Convention on the protection of the stratospheric ozone layer by limiting the manufacturing and use of chlorofluorocarbons (CFCs). Thus, an important beginning has been made. But it is only a beginning.

We cannot continue to subsidize or even tolerate practices that cause erosion, salinization, and groundwater contamination and depletion, or policies that make poor nations permanently dependent on the largesse of their rich neighbors. We must stop the destruction of habitats and the eradication of entire species. On the other hand, we cannot preach to the poor that they must remain so in order to protect the environment. They must have alternative and better means, not only of survival but of real progress.

No nation is free to damage its own domain if that damage affects others. A typical issue among many is whether the United States or China should have the right to build coal-fired power plants that will spew out oxides of carbon, nitrogen, and sulfur, acidifying the rain falling on its neighbors and contributing to the global greenhouse effect. Another example is the responsibility of the Soviet Union for the radioactive fallout of its nuclear accident affecting adjacent countries. Still another is the ability of Brazil to preserve the tropical rain forests of Amazonia, with their treasure trove of biological diversity, in view of that country's urgent need to develop its economy and alleviate the poverty of its people. The overarching issue is the duty of nations to each other and to the common global environment, and particularly the duty of the richer nations to help the poorer ones, in real terms rather than by saddling them with more and more burdensome debts.

Are the nations of the world ready at last to stop squandering the most precious of all resources—human effort—in the futile pursuit of sectarian military power or competitive economic advantage? It is sobering to realize that over the last two decades the total global investment in agricultural and environmental research and development has amounted to less than one percent of the total spent on armament. Is the community of nations now ready to begin applying the necessary means, heretofore wasted on weapons, toward the peaceful pursuit of education, population control, sustainable agricultural and economic development, and—last but certainly not least—restoration and protection of the environment?

As an agricultural and environmental scientist, I am convinced that we have the essential knowledge and capability to manage soil and water efficiently enough to feed all of humanity, even allowing for the unavoidable measure of expectable population growth. We may be daunted by the nature of the problems. But it behooves us to recall the ancient adage: "It is not for you alone to complete the task, but neither are you free to evade it." We have all come out of the earth, and are its children. The earth has always nurtured us, despite our scornful abuse, and we can no longer continue to behave as its ungrateful offspring. It is time for us, as *Homo sapiens curans,* to nurture the earth in return.

NOTES

1. PROLOGUE

1. The nomadic past of the Israelites must have remained forever etched in their collective memory, however, as only a pastoral people could have evoked the lore and imagery so poignantly expressed in the 23rd Psalm: "The Lord is my shepherd, I shall not want; He lays me down in green pastures, He leads me beside still waters. . . ."

2. For a description of that project in the Negev, see D. Hillel, *Negev: Land, Water, and Life in a Desert Environment* (1982).

3. The hazards of grand development programs that are conceived by great statesmen (who may understand much about world affairs but little about the environment) became all too apparent when our mission arrived in Burma and was shown the grassed flatlands (superficially reminiscent of the American prairie) where the missionary planners intended for us to raise wheat. I quickly discovered that the soil was a tropical laterite, with a stone-hard encrustation of iron oxide, totally unsuitable for cropping. We did, however, find other areas where a variety of crops could be grown, though not in accordance with the original plan.

4. These assignments were carried out on behalf of various international agencies, including the World Bank, the U.N. Food and Agriculture Organization, the U.S. Agency for International Development, the International Development Research Center, and the International Atomic Energy Agency.

1. The two accounts of creation are contrasted in detail in R. E. Friedmann, *Who Wrote the Bible* (1987).

2. The explicit name of God, considered ineffable in Hebrew but commonly rendered Jehovah in English, is not mentioned at all in the first chapter of Genesis and appears only in the second. The exact pronunciation of the Tetragrammaton *YHWH* is unknown (see Klein, 1987). The name probably derives from a combination of the three tenses of the verb "to be" in Hebrew: *haya* (was), *hoveh* (is), and *yihyeh* (will be), taken together to imply "The Eternal Being."

3. The verb used for Yahweh's forming of the earth-creature, *vayitzer*, is the specific verb for pottery (Isaiah 41:25; Jeremiah 18:4,6; Lamentations 4:2; and numerous other passages), and very different from the verb *bara* used to describe God's act of creating the heaven and the earth. That creation was a verbal command, whereas the forming of *ha'adam* is depicted as an active manual deed. Michelangelo missed the point in his portrayal of that act, showing God hovering over Adam and touching his hand delicately. Shaping him out of mud and blowing the breath of life into him would have been more in the spirit of Genesis 2:7.

4. "To serve and preserve it" is the author's translation of the Hebrew words *l'ovdah ul'shomrah* (Genesis 2:15), usually rendered "to dress it and keep it" (King James Version) or "to till it and keep it" (Revised Standard Version).

5. It seems strange that any religion could sanction man's right to unlimited mastery over nature. Since such a right necessarily includes the license to destroy, it implies God's abdication of responsibility or care for His own Creation, a notion that contradicts the very basis of most religions. Religious concern for God's Creation should therefore deny man's right to destroy any part of it. The reader might wonder why a book devoted to the science of the environment should digress into matters of religion. It does so because precepts of faith underlie human individual and collective actions, in the environmental as well as in other arenas. Moreover, science and religion (at least, monotheistic religion) share a fundamental tenet, namely the essential unity and interconnectedness of the universe.

6. The originally formed (Genesis 2:7) earth-being, *ha'adam* ("the" adam, implying "of the soil") is not specifically male. That creature was only later (Genesis 2:22) differentiated into man (*ish*) and woman (*isha*), with the respective names Adam and Eve.

7. An interesting exception is found in Egyptian mythology, in which the deity of the earth (Geb) was masculine, whereas that of the sky was feminine (often depicted as Geb's sister, Nut).

8. A fuller version of Chief Seattle's letter is given in Joseph Campbell, *The Power of Myth* (1988).

9. Prophet Smohalla's statement is quoted in J. Michell, *The Earth Spirit* (1975).
10. The word for rain in Hebrew is *geshem,* connoting "substantiation." To the ancients, who could not perceive that it is recycled water from the earth, rain represented the substantiation, or material manifestation, of God's grace from heaven. The Hebrew word for heaven, *shamayim,* can be separated into *sham-mayim,* which suggests "source of water."
11. In *Odyssey,* xi. 489.

3. THE FERTILE SUBSTRATE

1. William Blake was an artist as well as a poet. His view of the world, and the flow of spirit that animates it, is depicted in his tempera painting, "The Cycle of the Life of Man" (1821). In this striking work of art, men and women are shown rising out of the moist earth, to which, however, they remain rooted like plants.
2. Dr. Selman Waksman and his co-workers at Rutgers University, in the early 1940s.
3. Soil means different things to different people: to the houseplant grower, soil means ground peat; to the laundry worker, it means "dirt" to be removed from children's clothes; to the engineer, it is construction material for dams and roadbeds, on which and in which to place buildings; to the city dweller, it is dust; and to the farmer it is a seedbed and rootbed for crops. Finally, to the environmental scientist the soil is a natural body, engaged in dynamic interactions with the atmosphere above and the strata below, that serves as a host for innumerable living organisms.
4. For a detailed description of soil formation processes, *see* N. C. Brady, *The Nature and Properties of Soils* (1990).
5. For a detailed description of the nature and behavior of clay, *see* D. Hillel, *Fundamentals of Soil Physics* (1980).
6. A much grander generalization of the concept of composite life, viewing the entire biosphere as an integrated living entity, is described in James Lovelock, *The Ages of Gaia* (1988).
7. Brady (1990).

4. THE VITAL FLUID

1. Nor did people understand that all rain and snow originates as water that evaporated from the earth's surface and condensed in the upper atmosphere to form clouds. Water vapor was too thin and transparent to be considered substantial. The entire idea of the hydrological cycle, which many people today consider self-evident, was perceived only dimly before modern times and did not really take hold until the eighteenth century.

2. A description of the molecular structure and properties of water is given in D. Hillel, *Fundamentals of Soil Physics* (1980).

3. Vladimir Vernadsky's early insights are found in M. M. Kamshilov, *Evolution of the Biosphere* (Moscow: Mir Publishers, 1976).

4. The greenhouse effect is a natural phenomenon and basically a fortunate one. Thanks to it, the temperature of the earth's surface is some 30° Celsius above what it would otherwise have been. However, its artificial enhancement, caused by the progressive increase in the concentrations of certain radiatively active gases, may result in excessive blocking of outgoing heat and hence in a warming of the climate on a global scale. Among the gases involved are carbon dioxide, methane, nitrous oxide, chlorofluorocarbons, and tropospheric ozone. The effect of the latter is not to be confused with the different effect of stratospheric ozone in filtering out some of the incoming ultraviolet radiation.

5. THE DYNAMIC CYCLE

1. "The quality of mercy is not strained," wrote Shakespeare in *The Merchant of Venice*, "it droppeth as the gentle rain from heaven upon the place beneath. . . ." Little did Shakespeare know how destructive the "gentle rain" can be when "the place beneath" is not protected.

2. An elementary description of aquifers and of the entire hydrological cycle is given in the booklet by L. Leopold, *Water: A Primer* (1960).

3. Artesian aquifers may hold water under pressure, which may in places be great enough to cause water to rise in wells all the way to the surface without the need for pumping.

4. An alternative definition: The water table is the surface at which the pressure of water in the ground is equal to atmospheric pressure. Water in the saturated aquifer below the water table is at a pressure greater than atmospheric, whereas water in the (generally unsaturated) soil above the water table is at a sub-atmospheric pressure called "suction."

6. THE PRIMARY PRODUCERS

1. A basic description of the role of water in the life of plants is given in P. J. Kramer, *Water Relations of Plants* (1983).

2. A pictorial description of root growth was given by E. Epstein in *Scientific American* 228 (1973): 48–58.

3. A fundamental analysis of root–shoot growth relationships is given in M. G. Huck and D. Hillel, "A Model of Root Growth and Water Uptake Accounting for Photosynthesis, Respiration, Transpiration, and Soil Hydraulics," in *Advances in Irrigation,* Vol. 2 (1983), D. Hillel, ed.

1. An interesting analysis of the scientific method in relation to the physical and biological sciences was given by J. Bronowski in *A Sense of the Future* (1977). In it, he points out that whereas nineteenth-century science concentrated mainly on making specific measurements, twentieth-century science has concerned itself increasingly with finding general relationships.
2. All living things in the biosphere, continually needing to monitor and adjust to each other and to changing conditions, thus act in accord with the Red Queen's edict in *Alice in Wonderland:* "Here, you see, it takes all the running you can do, just to keep in the same place."
3. A very comprehensive review of the many changes in the land brought about by agriculture is given in M. G. Wolman and F. G. A. Fournier, eds., *Land Transformations in Agriculture* (1987).

8. HUMAN ORIGINS

1. A recent book on the evolution of the human race is by M. Harris, *Our Kind* (New York: Harper & Row, 1989). Articles on recent findings were written by Roger Lewin (*Science* 235 [1987]:969–71; and *Science* 239[1988]:1240–41) and by Elwyn L. Simons (*Science* 245[1989]:1343–50).
2. The Paleolithic and Neolithic Transformations are described in C. L. Redman, *The Rise of Civilization* (1978).
3. *See,* for instance, A. Goudie, *The Human Impact* (1982).
4. The extensive use of fire by the Australian aborigines is described in S. J. Hallam, *Fire and Hearth* (1979).

9. THE AGRICULTURAL TRANSFORMATION

1. *See,* for instance, M. N. Cohen, *The Food Crisis in Prehistory: Overpopulation and the Origins of Agriculture* (1977).
2. Rather thorough analyses of the factors underlying the advent of agriculture can be found in Hutchinson et al., *The Early History of Agriculture* (1977); and in the more recent book by Rindos, *The Origins of Agriculture: An Evolutionary Perspective* (1984).
3. C. L. Redman (1978).
4. An interesting summary of the Agricultural Transformation was given by J. Bronowski in his popular *The Ascent of Man* (Boston: Little, Brown & Co., 1973).
5. For a comprehensive description of plant domestication, *see,* for instance, Harlan (1975).
6. Goudie (1981).
7. Adams (1981).

10. EARLY FARMING IN THE NEAR EAST

1. Cohen (1977), Redman (1978).
2. *See* Zohary (1969, 1973), as well as Zohary and Hopf (1973).
3. Harlan (1975).
4. Zohary (1970).
5. Borowski (1987); Arnon and Raviv (1980).

11. SILT AND SALT IN MESOPOTAMIA

1. For a general description of the early civilizations of Mesopotamia and other parts of the Near East, *see* A. B. Knapp, *The History and Culture of Ancient Western Asia and Egypt* (1988).
2. The Mesopotamian stories of mankind's origin are also similar to the Biblical. In *Enuma Elish,* the Babylonian creation myth, man was created from the blood of the god Kingu mixed with dust from the ground.
3. Genesis 11:1–9.
4. An early analysis of the problems that beset the irrigation-based societies of southern Mesopotamia was given in 1958 by T. Jacobsen and R. M. Adams in, "Salt and Silt in Ancient Mesopotamian Agriculture," *Science* 128:1251–58.
5. *Ibidem.*
6. K. A. Wittfogel, "The Hydraulic Civilizations." In *Man's Role in Changing the Face of the Earth,* W. R. Thomas, ed., (Chicago: University of Chicago Press, 1956).
7. Jacobsen (1982).
8. Artzy and Hillel (1988).
9. *See,* for example, G. F. Dales, "The Decline of the Harappans," *Scientific American* 214(1966):92–100.

12. THE GIFTS OF THE NILE

1. Klein (1987).
2. *See The Oxford Dictionary of English Etymology* (C. T. Onions, ed. London: Oxford University Press, 1966).
3. Butzer (1976); James (1979).
4. A general pictorial description of Ancient Egypt is presented in C. Hobson, *The World of the Pharaohs* (New York: Thames and Hudson, 1987).
5. *See* B. G. Trigger, B. J. Kemp, D. O'Connor, and A. B. Lloyd, *Ancient Egypt: A Social History* (Cambridge: Cambridge University Press, 1983).
6. Wittfogel (1956).
7. *See* J. P. Oleson, *Greek and Roman Mechanical Water-Lifting Devices: The*

History of a Technology (Toronto: University of Toronto Press, 1984); as well Schioler, *Roman and Islamic Water-Lifting Wheels* (1973).

8. M. P. S. Girard, "Mémoire sur l'agriculture, l'industrie et le commerce de l'Égypte," in *Description de l'Égypte: État Moderne,* Vol. II (1812):491–714. Paris: Imperial Printers.

13. HUSBANDRY OF THE RAIN-FED UPLANDS

1. Thirgood (1981).
2. An early description of land degradation in the Mediterranean region and elsewhere was given by Lowdermilk (1953).
3. Hillel (1980b).
4. Carter and Hale (1974).
5. Borowski (1987).
6. One of the most remarkable hydraulic works in ancient Israel is found in Jerusalem, where the pulsating karstic spring of Gihon (meaning "gusher") was diverted via a tunnel from outside the walls into the fortified city. The project was undertaken by King Hezekiah (ca. 700 B.C.E.) in anticipation of the siege by Assyrian King Sennacherib. In haste to complete the work before the enemy arrived, the workers excavated the 530-meter tunnel from both ends. Where they met inside the mountain, they left an inscription expressing joy at the successful completion of their difficult task. Indeed, this diversion saved Jerusalem from succumbing to King Sennacherib, who some time later lifted the siege and returned to Assyria. Adventurous visitors can wade through the tunnel today. So important was the spring of Gihon, being the main source of water in the immediate environs of ancient Jerusalem (and probably the reason for the city's original location), that it was selected as the site for the coronation of Solomon, the first king crowned in Jerusalem.
7. Deuteronomy 15:1.
8. Isaiah, 5.
9. Hillel (1982).
10. Carter and Hale (1974).

14. THE DESERT REJOICED

1. Jeremiah 2:2.
2. For a comprehensive description of the Negev, *see* D. Hillel, *Negev: Land, Water, and Life in a Desert Environment* (1982).
3. A. Negev, "The Nabateans in the Negev," in *The Land of the Negev* (Tel Aviv: Ministry of Defense Publications, 1979).
4. Hillel (1970).
5. *See* M. Evenari, L. Shanan, and N. Tadmor, *The Negev: The Challenge of a Desert,* 2nd ed. (Cambridge, Mass.: Harvard University Press, 1982).

6. D. Hillel and N. Tadmor, "Water Regime and Vegetation in the Negev Highlands of Israel," *Ecology* 43 (1962):33–41.
7. Hillel (1982).
8. McGregor (1982).
9. *See* R. H. and F. C. Lister, *Chaco Canyon* (Albuquerque: University of New Mexico Press, 1981). Also, *see* S. H. Lekson, T. C. Windes, J. R. Stein, and W. J. Judge, "The Chaco Canyon Community," *Scientific American* 259 (1988):100–109.

15. TAPPING THE UNDERGROUND WATERS

1. Genesis 24:16.
2. Genesis 21:19.
3. There is, however, some evidence that *qanat* building was started in Armenia, rather than in Persia. Assyrian King Sargon II (721–705 B.C.E.), boasting of his conquest of a town named Ulhu in Urartu (present-day Armenia), described how the king of that town had "revealed the water outlets . . . water of abundance he caused to flow . . . countless ditches he led out from its interior . . . and he irrigated the fields." *See* J. Lassoe, "The Irrigation System at Ulhu," *Journal of Cuneiform Studies* 5 (1951):21–32.

16. FARMING THE WETLANDS OF MESOAMERICA

1. *See*, for instance, the recent comprehensive article by W. E. Garrett, "La Ruta Maya," *National Geographic* 176:424–79.
2. An impressive report on these systems was published by C. Plazas and A. M. Falchetti de Saenz, *Asentamientos Prehispánicos en el Bajo Río San Jorge* (Bogotá, Colombia: Fundación de Investigaciones Arqueológicas Nacionales, 1981). *See also* W. M. Denevan, K. Mathewson, and G. Knapp, eds. (1987), *Pre-Hispanic Agricultural Fields in the Andean Region* (Proceedings of the International Congress of Americanists, Bogotá, Colombia, 1985).

17. THE ADVENT OF CHEMICAL FERTILIZERS

1. Brady (1990).
2. Wild (1988).
3. That unfortunate principle, incidentally, is exemplified in the life of Fritz Haber himself. Having invented a process of such great value to soil fertility (an invention for which he was awarded the Nobel Prize), he then went on to apply the same method to help Germany manufacture

explosives for use in World War I, and even played a leading part in developing poison gas as a weapon. All his patriotic zeal for the German Fatherland, ironically, availed him naught when, in 1933, he was hounded out of Germany with the epithet, "The Jew Haber." He died in exile the same year.

18. SALINE SEEPS IN AUSTRALIA AND NORTH AMERICA

1. Several important articles analyzing the saline seep phenomenon, particularly in Australia, are included in J. W. Holmes and T. Talsma, eds., *Land and Stream Salinity* (Amsterdam: Elsevier, 1981).
2. The saline seep phenomenon in North America is reviewed in an article by M. R. Miller et al., "Saline Seep Development and Control in the North American Great Plains," in Holmes and Talsma, eds., *Land and Stream Salinity;* as well as in an article by R. B. Daniels, "Saline Seeps in the Northern Great Plains of the USA and the Southern Prairies of Canada," in M. G. Wolman and F. G. A. Fournier, eds., *Land Transformation in Agriculture* (New York: John Wiley & Sons, 1987).

19. THE PROMISE AND PERIL OF IRRIGATION

1. Hillel (1987).
2. A comprehensive survey of irrigation development around the world is given in Hitoshi Fukuda, *Irrigation in the World* (Tokyo: University of Tokyo Press, 1976).
3. Farr and Henderson (1986).
4. Excellent reviews of the Indus Basin drainage problem were given by W. Barber, *The Rising Water Table and Development of Water Logging in Northwestern India* (presented to the South Asia Department of the World Bank, Washington, D.C., 1985); and by L. Shanan in Wolman and Fournier, eds., *Land Transformation in Agriculture.*
5. In the New Lands of Egypt, where I worked as a consultant in the effort to improve irrigation efficiency, I witnessed a rather depressing scene of misapplied technology. Huge center-pivot irrigation rigs, installed just a few years before in the hope of achieving instantaneous modernization, stood motionless and rusting under the sun. Intended to circle continuously and sprinkle water onto lush crops, as they do in the Great Plains of the western United States, these imported supermachines are thoroughly alien in the context of Egyptian irrigation. One broken cog is all it takes to paralyze such an import-dependent behemoth and turn it from a life-giver into a grotesque "white elephant." Had the same investment been applied to developing smaller units made

locally and adapted to the scale and character of irrigation in Egypt, and had the people operating them been given direct ownership or effective incentives to maintain them, the scheme would probably have worked much better.

6. Quite a different problem now threatens the soils of Egypt. The country's rapid urbanization requires bricks, and these are made of the clayey soil of the Nile's floodplain. The mining of this precious soil is now being carried out on a large scale. Many farmers are unable to resist the lucrative sums offered them for their small plots of land, and are enticed to sell their birthright for a mess of pottage. They end up bereft of the real resource that had sustained their ancestors for countless generations. The soil is then dug up and carted away to brick factories, leaving pits and hollows that can never be farmed again. Though the practice is illegal, it continues nevertheless. Egypt's most precious resource is thereby subjected to instant erosion, which, in the absence of the annual replenishment of silt by the Nile, is truly irreversible.

One is reminded of the exchange between Wang Lung and his sons in Pearl Buck's *The Good Earth.* The sons say: "This field we shall sell and that one, and we will divide the money between us," to which the old man replies: "Out of the land we came, and into it we must go— and if you will hold the land you can live—no one can rob you of your land. . . . If you sell the land, it is the end."

7. J. Quiggin, "Murray River Salinity—An Illustrative Model," *American Journal of Agricultural Economics* 70 (1988):635–45.

8. P. P. Micklin, "Desiccation of the Aral Sea: A Water Management Disaster in the Soviet Union," *Science* 241 (1988):1170–76. A pictorial account of the Aral is given in the recent article by W. S. Ellis, "A Soviet Sea Lies Dying," *National Geographic* 177 (1990):73–92.

9. As the Aral Sea diminishes, one recalls sadly the formerly glorious lake celebrated by Matthew Arnold in his epic poem, *Sohrab and Rustum:*

The shorn and parcelled Oxus strains along
Through beds of sand and matted rushy isles—
Oxus, forgetting the bright speed he had
In his high mountain cradle in Pamere,
A foiled circuitous wanderer; till at last
The longed-for dash of waves is heard, and wide
His luminous home of waters opens, bright
And tranquil, from whose floor the new-bathed stars
Emerge, and shine upon the Aral Sea.

10. The hubris of central state planners in the Soviet Union, impelled by Stalin's vision to "transform nature," was expressed in 1926 by Zazubrin: "Let the fragile green breast of Siberia be dressed in the cement armor

of cities, armed with the stone muzzles of factory chimneys, and girded with iron belts of railroads. Let the taiga be burned and felled, let the steppes be trampled. Let this be, and so it will be inevitably. Only in cement and iron can the fraternal union of all peoples, the iron brotherhood of all mankind, be forged."

Tolstoy anticipated this fallacy with striking accuracy in his "Ballad with a Bias:"

They want to flatten the whole world
And thus introduce equality,
They want to spoil everything
For the common good.

11. Hillel (1987).

20. ACCELERATED EROSION

1. *Soils and Men: 1938 Yearbook of Agriculture,* U.S. Department of Agriculture, Washington, D.C.
2. Tank (1983).
3. An important collection of papers reviewing the past and present state of erosion and conservation was published in Douglas Helms and Susan L. Flader, eds., *The History of Soil and Water Conservation* (published by the University of California Press for The Agricultural History Society, Washington, D.C., 1985).
4. *See,* for example, Paul Bonnifield, *The Dust Bowl: Men, Dirt, and Depression* (1979).
5. ". . . the dust lifted up out of the fields," wrote Steinbeck in his opening chapter to *The Grapes of Wrath,* "and drove gray plumes into the air like sluggish smoke. . . . The finest dust did not settle back to earth, but disappeared into the darkening sky. . . . The air and sky darkened and through them the sun shone redly, and there was a raw sting in the air . . . and the corn fought the wind with its weakened leaves until the roots were freed by the prying wind and then each stalk settled wearily . . . and the wind cried and whimpered over the fallen corn."
6. *See,* for example, George R. Foster, ed., *Soil Erosion: Prediction and Control* (Ankeny, Ohio: Soil Conservation Society of America, 1977).
7. Hudson (1971).

21. THE "SORROW OF CHINA"

1. A striking pictorial description of this soil is given in Wang Yon-yan and Zhang Zong-hu, eds., *Loess in China* (Shaanxi People's Art Publishing House, 1980).

2. Zhang Zezhen and Deng Shangshi, "The Development of Irrigation in China," *Water International* 12 (1987):46–52.
3. The classic book by Franklin H. King describes how farmers in East Asia cultivated land for some 4,000 years without depleting the fertility of their soil. *See Farmers of Forty Centuries: Permanent Agriculture in China, Korea, and Japan* (Madison, Wis., 1911; New York: Harcourt, Brace, 1927).

22. DEFORESTING THE EARTH

1. Goudie (1982).
2. Thirgood (1981).
3. David Grigg, "The Industrial Revolution and Land Transformation," in Wolman and Fournier, eds., *Land Transformation in Agriculture.*
4. *See* the booklet by Sandra Postel and Lori Heise, *Reforesting the Earth,* Worldwatch Paper 83, Worldwatch Institute, Washington, D.C., 1988. *See also* Robert Repetto and Malcolm Gillis, eds., *Public Policies and the Misuse of Forest Resources* (Cambridge: Cambridge University Press, 1988).
5. A recent and comprehensive book on the Amazonian forest is by Susanna Hecht and Alexander Cockburn, *The Fate of the Forest: Developers, Destroyers and Defenders of the Amazon* (London: Verso, 1989).
6. Hillel and Rosenzweig (1989).
7. *See,* for example, R. B. and R. A. Boyle, *Acid Rain* (New York: Schocken Books, 1983), as well as the booklet by Sandra Postel, *Air Pollution, Acid Rain, and the Future of Forests,* Worldwatch Paper 58, Worldwatch Institute, Washington, D.C., 1984.

23. MAN-MADE DESERTS

1. Dregne (1983).
2. J. E. Gorse and E. R. Steeds, *Desertification in the Sahelian and Sudanian Zones of West Africa,* World Bank Technical Paper No. 61, Washington, D.C., 1987.
3. S. E. Nicholson, *The Sahel: A Climatic Perspective,* Report to the Organization for Economic Cooperation and Development (OECD), Washington, D.C., 1982.
4. *See,* for instance, the booklet edited by Francois Falloux and Aleki Mukendi, *Desertification Control and Renewable Resource Management in the Sahelian and Sudanian Zones of West Africa,* World Bank Technical Paper No. 70, Washington, D.C. *See also* the paper by Ridley Nelson, *Dryland Management: The "Desertification" Problem,* Environment Department Working Paper No. 8, World Bank, 1988.

5. L. M. Druyan, "Advances in the Study of Sub-Saharan Drought," *International Journal of Climatology* 9 (1989):77–90.
6. Important among the radiatively active gases of increasing concentration are carbon dioxide (released by the burning of fossil fuels, the clearing of forests, and the decomposition of organic matter in newly cultivated soils), methane (released from wetlands and landfills, as well as by termites and bovines), ozone (formed photochemically in the lower atmosphere by the reaction of various emissions with sunlight), nitrous oxide (released from the soil under partially anaerobic conditions), and finally, a family of synthetic gases called chlorofluorocarbons (CFCs). The latter are artificial gases blithely insinuated into the atmosphere by our innovative chemical industry during the last 50 years, to serve as refrigerants, foaming agents, and spray-propellants. The CFCs not only contribute to the greenhouse effect, but also, as they diffuse into the stratosphere, act to destroy the ozone layer that helps to block some of the incoming mutagenic ultraviolet radiation.

24. THE PLIGHT OF AFRICA

1. L. Timberlake, *Africa in Crisis: The Causes, the Cures of Environmental Bankruptcy* (1985).
2. L. R. Brown and E. C. Wolf, *Reversing Africa's Decline,* Worldwatch Paper 65, Worldwatch Institute, Washington, D.C., 1985.
3. The scene brings to mind the lines from T. S. Eliot's *The Waste Land:*

What are the roots that clutch, what branches grow
Out of this stony rubbish? Son of man,
You cannot say, or guess, for you know only
A heap of broken images, where the sun beats
And the dead tree gives no shelter, the cricket no relief,
And the dry stone no sound of water. . . .
And I will show you something different from either
Your shadow at morning striding behind you
Or your shadow at evening rising to meet you;
I will show you fear in a handful of dust.

4. See *Atlas of African Agriculture,* U.N. Food and Agriculture Organization, Rome, Italy, 1986.
5. Lal (1987).
6. Sanchez (1976).
7. R. Lal, "Soil Degradation and the Future of Agriculture in Sub-Saharan Africa," *Journal of Soil and Water Conservation* 43 (1988):444–51.
8. Brown and Wolf (ibid.).

NOTES   1. E. Maltby, *Waterlogged Wealth: Why Waste the World's Wet Places?* (1986).
2. *See,* for example, the comprehensive article by K. Ruddle, "The Impact of Wetland Reclamation," in Wolman and Fournier (1987).

26. SWEET WATER AND BITTER

1. *See* article by J. W. Maurits la Riviere, "Threats to the World's Water," *Scientific American* 261 (1989):80–94.
2. Sandra Postel, *Water: Rethinking Management in an Age of Scarcity,* (Worldwatch Paper 62, Worldwatch Institute, Washington, D.C., 1984.
3. *See,* for example, the article by C. Cheverry, F. Fournier, and S. Henin, "Criteria for Observing and Measuring Changes Associated with Land Transformations," in Wolman and Fournier (1987).
4. G. Gee and D. Hillel, "Groundwater Recharge in Arid Regions," *Hydrological Processes* 2 (1988):255–66.
5. A comprehensive report on the Ogallala Aquifer was prepared for the U.S. Department of Commerce by the High Plains Associates (a consortium of engineering firms), under the title *Six-State High Plains Ogallala Aquifer Regional Resources Study,* Austin, Texas, 1982.
6. S. Postel, *Conserving Water: The Untapped Alternative,* Worldwatch Paper 67, Worldwatch Institute, Washington, D.C., 1985.
7. M. Reisner, 1986.
8. C. Cheverry et al., in Wolman and Fournier (1987).
9. T. Naff and R. C. Matson (1984).
10. In the words of the Koran: "It is He who has released two bodies of flowing water: One palatable and sweet, the other salty and bitter; yet He has separated between them. . . . It is He who has created man from water" (Sura XXV: 53–54).

27. WATER MANAGEMENT IN ISRAEL

1. S. Arlosoroff, "Water Resources Development and Management in Israel," *Kidma* 5 (1978):4–10.
2. A. Wiener, "Water Planning at Home and Abroad," *Kidma* 3 (1976):36–39.
3. E. Rawitz and D. Hillel, "Progress and Problems of Drip Irrigation in Israel," *Proceedings International Conference on Drip Irrigation,* San Diego, California, (1975).

1. Regenstein (1982).
2. *See* E. P. Jorgensen, ed., *The Poisoned Well* (Washington, D.C.: Island Press, 1989).
3. S. Postel, *Defusing the Toxics Threat: Controlling Pesticides and Industrial Waste,* Worldwatch Paper 79, Worldwatch Institute, Washington, D.C., 1987.
4. The toxic waste problem brings to mind the lines from William Empson's "Missing Dates": "Slowly the poison the whole stream fills. It is not the effort nor the failure tires. The waste remains, the waste remains and kills."

29. A GLOBAL ACCOUNTING

1. N. Keyfitz (1989), "The Growing Human Population," *Scientific American* 261:118–26.
2. A. C. Kelly, "Economic Consequences of Population Change in the Third World," *Journal of Economic Literature* 26 (1988): 1685–1728.
3. *See* V. Smil, P. Nachman, and T. V. Long, *Energy Analysis and Agriculture: An Application to U.S. Corn Production* (Boulder, Colo.: Westview Press, 1983).
4. *See,* for example, the recent report of the National Research Council, *Alternative Agriculture* (Washington, D.C.: National Academy Press, 1989); and Clive A. Edwards, Rattan Lal, Patrick Madden, Robert H. Miller, and Gar House, eds., *Sustainable Agricultural Systems* (Ankeny, Ohio: Soil and Water Conservation Society, 1990).
5. P. M. Vitousek, P. R. Ehrlich, A. H. Ehrlich, and P. A. Matson, "Human Appropriation of the Products of Photosynthesis," *Bioscience* 36 (1986): 402–11.
6. In a speech, "The World Bank's Challenge: Balancing Economic Need with Environmental Protection," delivered to the World Wildlife Fund, London, March 3, 1988.

SELECTED
BIBLIOGRAPHY

Adams, Robert McC. 1981. *Heartland of Cities: Surveys of Ancient Settlement and Land Use on the Central Floodplain of the Euphrates*. Chicago: University of Chicago Press.

Ahmad, Nazir, and Ghulam Rasul Chaudhry. 1988. *Irrigated Agriculture of Pakistan*. Lahore, Pakistan: Shahzad Nazir.

Anon. 1974. *Productive Agriculture and a Quality Environment*. Washington, D.C.: National Academy of Sciences.

———. 1982. *Soil Erosion and Conservation in the Tropics*. Madison, Wis.: American Society of Agronomy, Special Publ. No. 43.

———. 1983. *Environmental Change in the West African Sahel*. Washington, D.C.: National Academy Press.

———. 1986. *The Encroaching Desert: The Consequences of Human Failure*. London: Zed Books.

———. 1974. *More Water for Arid Lands: Promising Technologies and Research Opportunities*. Washington, D.C.: National Academy of Sciences.

———. 1986. *Sahel Report: A Long-Term Strategy for Environmental Rehabilitation*. Gland, Switzerland: IUCN (International Union for the Conservation of Nature and Natural Resources).

Artzy, Michal, and Daniel Hillel. 1988. "A Defense of the Theory of Progressive Soil Salinization in Ancient Southern Mesopotamia." *Geoarchaeology* Vol 3, pp. 235–38.

Baker, Herbert G. 1965. *Plants and Civilization*. Belmont, Cal.: Wadsworth.

Barreiro, Jose. 1988. "The Sioux Look to the Earth." *Daybreak* Vol. 2, pp. 10–15.

Barrow, Chris. 1987. *Water Resources and Agricultural Development in the Tropics.* Harlow, Essex, England: Longman.

Barth, Michael C., and James G. Titus. 1984. *Greenhouse Effect and Sea Level Rise.* New York: Van Nostrand Reinhold.

Bear, Firman E. 1986. *Earth: The Stuff of Life (2d. Ed.).* Norman, Okla.: University of Oklahoma Press.

Beasley, R. P., James M. Gregory, and Thomas R. McCarty. 1984. *Erosion and Sediment Pollution Control.* Ames, Iowa: Iowa University Press.

Berry, Thomas. 1988. *The Dream of the Earth.* San Francisco: Sierra Club Books.

Berry, Wendell. 1977. *The Unsettling of America: Culture and Agriculture.* San Francisco: Sierra Club Books.

Black, John. 1970. *The Dominion of Man.* Edinburgh: Edinburgh University Press.

Bonnifield, Paul. 1979. *The Dust Bowl: Men, Dirt, and Depression.* Albuquerque, N.M.: University of New Mexico Press.

Borowski, Oded. 1987. *Agriculture in Iron Age Israel.* Winona Lake, Ind.: Eisenbrauns.

Brady, Nyle C. 1990. *The Nature and Properties of Soils.* New York: Macmillan.

Bras, Rafael L. 1990. *Hydrology: An Introduction to Hydrologic Science.* Reading, Mass.: Addison Wesley.

Briggs, David, and Frank Courtney. 1985. Harlow, Essex, England: Longman.

Bronowski, J. 1977. *A Sense of the Future.* Cambridge, Mass.: MIT Press.

Brouwer, C., A. Goffeau, and M. Heibloem. 1985. *Introduction to Irrigation.* Rome: U.N. Food and Agriculture Organization.

Brown, M. T., and H. T. Odum. 1981. *Research Needs for a Basic Science of the System of Humanity and Nature, and Appropriate Technology for the Future.* Gainesville, Fla.: University of Florida.

Brundtland, Gro Harlem, Chairman. 1987. *Our Common Future: Report of the World Commission on Environment and Development.* Oxford: Oxford University Press.

Butzer, Karl W. 1976. *Early Hydraulic Civilization in Egypt: A Study in Cultural Ecology.* Chicago: University of Chicago Press.

———. 1971. *Environment and Archeology: An Ecological Approach to Prehistory.* Chicago: Aldine-Atherton.

Caldwell, John C., and Pat Caldwell. 1990. "High Fertility in Sub-Saharan Africa." *Scientific American* 262:118–125.

Campbell, Joseph. 1988. *The Power of Myth*. New York: Doubleday.

Carpenter, Richard A., ed. 1983. *Natural Systems for Development*. New York: Macmillan.

Carter, Vernon Gill, and Tom Hale. 1974. *Topsoil and Civilization*. Norman, Okla.: University of Oklahoma Press.

Child, R. Dennis. 1984. *Arid and Semiarid Lands: Sustainable Use and Management in Developing Countries*. Morrilton, Ark.: Winrock International.

Chiras, Daniel D. 1985. *Environmental Science: A Framework for Decision Making*. Menlo Park, Cal.: Benjamin/Cummings.

Clark, W. C., and R. E. Munn. 1986. *Sustainable Development of the Biosphere*. Cambridge: Cambridge University Press.

Clawson, Marion. 1964. *Man and Land in the United States*. Lincoln, Neb.: University of Nebraska Press.

Cohen, Nathan Mark. 1977. *The Food Crisis in Prehistory: Overpopulation and the Origins of Agriculture*. New Haven, Conn.: Yale University Press.

Cook, R. L. 1962. *Soil Management for Conservation and Production*. New York: John Wiley & Sons.

Cook, Ray L., and Boyd G. Ellis. 1987. *Soil Management: A World View of Conservation and Production*. New York: John Wiley & Sons.

Cuenca, Richard H. 1989. *Irrigation System Design*. Englewood Cliffs, N.J.: Prentice Hall.

Daly, Herman E., Editor. 1980. *Economics, Ecology, Ethics: Essays Toward a Steady-State Economy*. San Francisco: W.H. Freeman.

De Marsily, Ghislain. 1986. *Quantitative Hydrogeology*. San Diego, Cal.: Academic Press.

De Candolle, Alphonse. 1967. *Origin of Cultivated Plants*. New York: Harper.

De Beus, J. G. 1985. *Shall We Make the Year 2000?* London: Sidgwick & Jackson.

Dregne, H. E. 1983. *Desertification of Arid Lands*. Chur, Switzerland: Harwood Academic.

Dubos, Rene. 1980. *The Wooing of Earth*. New York: Charles Scribner's Sons.

Dunne, Thomas, and Luna B. Leopold. 1978. *Water in Environmental Planning*. San Francisco: W. H. Freeman.

Eckholm, Erik P. 1982. *Down to Earth: Environment and Human Need*. New York: W. W. Norton.

304

Eckholm, Erik P. 1976. *Losing Ground: Environmental Stress and World Food Prospects.* New York: W. W. Norton.

Fanning, Delvin S., and Mary C. B. Fanning. 1989. *Soils: Morphology, Genesis, and Classification.* New York: John Wiley & Sons.

Farr, E., and W. C. Henderson. 1986. *Land Drainage.* Harlow, Essex, England: Longman.

Feliks, Jehuda. 1963. *Agriculture in Palestine in the Period of the Mishna and Talmud.* Jerusalem: Magnes Press.

Finegan, Jack. 1979. *Archaeological History of the Ancient Middle East.* New York: Dorset Press.

Finkel, Herman J. 1986. *Semiarid Soil and Water Conservation.* Boca Raton, Fla.: CRC Press.

Finkelstein, Israel. 1988. *The Archaeology of the Israelite Settlement.* Jerusalem: Israel Exploration Society.

Friedman, Richard Elliott. 1987. *Who Wrote the Bible.* New York: Summit Books.

Fukuda, Hitoshi. 1976. *Irrigation in the World: Comparative Developments.* Tokyo: University of Tokyo Press.

Fuller, R. Buckminster. 1971. *Operating Manual for Spaceship Earth.* New York: E. P. Dutton.

Furon, Raymond. 1967. *The Problem of Water: A World Study.* New York: American Elsevier.

Galbraith, John Kenneth. 1977. *The Age of Uncertainty.* London: British Broadcasting Corporation.

Garraty, John A., and Peter Gay. 1987. *The Columbia History of the World.* New York: Harper & Row.

Gerasimov, I. P. (Ed.) 1975. *Man, Society and the Environment.* Moscow: Progress Publishers.

Goudie, Andrew. 1982. *The Human Impact: Man's Role in Environmental Change.* Cambridge, Mass.: MIT Press.

Gradus, Yehuda, ed. 1985. *Desert Development: Man and Technology in Sparselands.* Dordrecht, Netherlands: D. Reidel Publishing Co.

Green, Donald E. 1973. *Land of the Underground Rain: Irrigation on The Texas High Plains.* Austin, Tex.: University of Texas Press.

Gregorios, Paulos Mar. 1987. *The Human Presence: Ecological Spirituality and the Age of the Spirit.* Warwick, N.Y.: Amith House.

Gribbin, John. 1982. *Future Weather and the Greenhouse Effect.* New York: Dell Publishing Co.

Hallam, Sylvia J. 1979. *Fire and Hearth: A Study of Aboriginal Usage and European Usurpation in South-Western Australia.* Canberra: Australian Institute of Aboriginal Studies.

Hallsworth, E. G. 1987. *Anatomy, Physiology and Psychology of Erosion.* New York: John Wiley & Sons.

Hamilton, Michael (Ed.). 1970. *This Little Planet.* New York: Charles Scribner's Sons.

Harlan, Jack R. 1975. *Crops and Man.* Madison, Wis.: American Society of Agronomy.

Harrison, Edward. 1985. *Masks of the Universe.* New York: Macmillan.

Hauser, Philip M. 1979. *World Population and Development: Challenges and Prospects.* Syracuse, N.Y.: Syracuse University Press.

Hawley, Amos H. 1975. *Man and Environment.* New York: Franklin Watts, Inc.

Hillel, Daniel. 1970. "Artificial Inducement of Runoff as a Potential Source of Water in Arid Lands." In McGinnies, W. G., ed. *Food, Fiber, and the Arid Lands.* Tucson, Ariz.: University of Arizona Press.

————. 1971. *Soil and Water: Physical Principles and Processes.* New York, N.Y.: Academic Press.

————. 1972. *Optimizing the Soil Physical Environment Toward Greater Crop Yields.* New York: Academic Press.

————. 1976. "Soil Management in Arid Regions." In *Yearbook of Science and Technology.* New York: McGraw-Hill.

————. 1976. "A New Method of Water Conservation in Arid Zone Soils." In Mundlak, Y., and S.F. Singer, Editors. *Arid Zone Development.* Cambridge, Mass.: Ballinger (Published for the American Academy of Arts and Sciences).

————. 1979. "Irrigation and Crop Response." In *The Role of Soil Physics in Maintaining the Productivity of Tropical Soils.* Ibadan, Nigeria: International Institute for Tropical Agriculture.

————. 1980. *Fundamentals of Soil Physics.* San Diego, Calif.: Academic Press.

————. 1980. *Applications of Soil Physics.* San Diego, Calif.: Academic Press.

————. 1982. *Negev: Land, Water, and Life in a Desert Environment.* New York, N.Y.: Praeger.

————. 1987. *The Efficient Use of Water in Irrigation: Principles and Practices for Improving Irrigation in Arid and Semiarid Regions.* Washington, D.C.: World Bank.

————. 1988. "The Fate of Organic Contaminants in Soils." In *Petroleum in the Soil Environment.* Chelsea, Mich.: Lewis Publishers.

———. 1990. "The Role of Irrigation in Agricultural Systems." In *Irrigation of Agricultural Crops*. Madison, Wis.: American Society of Agronomy.

———. ed. 1982. *Advances in Irrigation, Vol. I*
1983. *Advances in Irrigation, Vol. II*
1985. *Advances in Irrigation, Vol. III*
1987. *Advances in Irrigation, Vol. IV* San Diego, Cal.: Academic Press.

Hillel, Daniel and Cynthia Rosenzweig. 1989. *The Greenhouse Effect and Its Implications Regarding Global Agriculture*. Amherst, Mass.: College of Food and Natural Resources, University of Massachusetts.

Holmes, J. W., and T. Talsma, Editors. *Land and Stream Salinity*. Amsterdam: Elsevier.

Holy, Milos. 1980. *Erosion and Environment*. Oxford: Pergamon Press.

Howard, Robert West. 1985. *The Vanishing Land*. New York: Ballantine Books.

Hudson, Norman. 1971. *Soil Conservation*. London: Batsford.

Hughes, J. Donald. 1975. *Ecology in Ancient Civilizations*. Albuquerque: University of New Mexico Press.

Hutchinson, Joseph, Grahame Clark, E. M. Jope, and R. Riley. 1977. *The Early History of Agriculture*. Oxford: Oxford University Press.

Hyams, Edward. 1976. *Soil and Civilization*. New York: Harper & Row.

Jackson, Wes, Wendell Berry, and Bruce Colman, eds. *Meeting the Expectations of the Land: Essays in Sustainable Agriculture and Stewardship*. 1984. San Francisco: North Point Press.

Jackson, I. J. 1989. *Climate, Water, and Agriculture in the Tropics*. Harlow, Essex, England: Longman.

Jacobsen, Thorkild. 1982. *Salinity and Irrigation Agriculture in Antiquity*. Malibu, Cal.: Undena Publications.

James, L. Douglas, ed. 1974. *Man and Water: The Social Sciences in Management of Water Resources*. Lexington, Ken.: University Press of Kentucky.

James, T. G. H. 1979. *An Introduction to Ancient Egypt*. New York: Farrar Straus Giroux.

Jordan, Wayne R., ed. 1987. *Water and Water Policy in World Food Supplies*. College Station, Tex.: Texas A&M University Press.

Kellogg, Charles E. 1975. *Agricultural Development: Soil, Food, People, Work*. Madison, Wis.: Soil Science Society of America.

Khouri, Rami G. 1981. *The Jordan Valley: Life and Society Below Sea Level*. London: Longman.

Klein, Ernest. 1987. *A Comprehensive Etymological Dictionary of the Hebrew Language*. Jerusalem: Carta.

Knapp, A. Bernard. 1988. *The History and Culture of Ancient Western Asia.* Chicago: Dorsey Press.

Komarov, Boris. 1980. *The Destruction of Nature in the Soviet Union.* White Plains, N.Y.: M. E. Sharpe.

Kramer, Paul, J. 1983. *Water Relations of Plants.* San Diego, Cal.: Academic Press.

Kuenen, P. H. 1963. *Realms of Water.* New York: John Wiley & Sons.

Lal, R. 1987. *Tropical Ecology and Physical Edaphology.* New York: John Wiley & Sons.

Lal, R., P. A. Sanchez, and R. W. Cummings, eds. 1986. *Land Clearing and Development in the Tropics.* Rotterdam: Balkema.

Leopold, Luna B. 1974. *Water: A Primer.* San Francisco: W. H. Freeman.

Lister, Robert H., and Florence C. Lister. 1981. *Chaco Canyon: Archaeology and Archaeologists.* Albuquerque, N.M.: University of New Mexico Press.

Lovelock, James. 1988. *The Ages of Gaia.* New York: W. W. Norton.

Lowdermilk, W. C. 1953. *Conquest of the Land Through 7,000 Years.* Washington, D.C.: U.S. Department of Agriculture, Soil Conservation Service, Bull. 99.

Mahar, Dennis J. 1989. *Government Policies and Deforestation in Brazil's Amazon Region.* Washington, D.C.: World Bank.

Maltby, Edward. 1986. *Waterlogged Wealth: Why Waste the World's Wet Places?* London: International Institute for Environment and Development.

Manger, Leif O. 1981. *The Sand Swallows Our Land.* Bergen, Norway: University of Bergen.

Mather, John. 1984. *Water Resources: Distribution, Use, and Management.* New York: John Wiley & Sons.

McCool, Donald K., ed. 1985. *Erosion and Soil Productivity.* St. Joseph, Mich.: American Society of Agricultural Engineers.

McGregor, John C. 1982. *Southwestern Archaeology.* Urbana, Ill.: University of Illinois Press.

McKibben, Bill. 1989. *The End of Nature.* New York: Random House.

McPhee, John. 1989. *The Control of Nature.* New York: Farrar Straus Giroux.

Mellor, John W., and Frank Z. Riely. 1989. "Expanding the Green Revolution." *Issues in Science and Technology VI,* pp. 66–74.

Michell, John. 1975. *The Earth Spirit: Its Ways, Shrines, and Mysteries.* New York: Thames and Hudson.

Morgan, R. P. C. 1986. *Soil Erosion and Conservation.* Harlow, Essex, England: Longman.

Naff, Thomas, and Ruth C. Matson. 1984. *Water in the Middle East: Conflict or Cooperation?* Boulder, Colo.: Westview Press.

O'Mara, Gerald T., ed. 1988. *Efficiency in Irrigation: The Conjunctive Use of Surface and Groundwater Resources.* Washington, D.C.: World Bank.

Page, G. William. 1987. *Planning for Groundwater Protection.* San Diego, Cal.: Academic Press.

Parfit, Michael. 1989. "Facing up to Reality in the Amazon." *Smithsonian* Vol. 20, pp. 58–77.

Parker, Ronald B. 1984. *Inscrutable Earth.* New York: Charles Scribner's Sons.

Pecsi, Marton, ed. 1987. *Loess and Environment.* Cremlingen-Destedt, West Germany: Catena Verlag.

Pittman, Nancy P., ed. 1988. *From the Land.* Washington, D.C.: Island Press.

Poincelot, Raymond P. 1986. *Toward a More Sustainable Agriculture.* Westport, Conn.: AVI Publishing Co.

Postel, Sandra. 1989. *Water for Agriculture: Facing the Limits.* Washington, D.C.: Worldwatch Institute.

Powledge, Fred. 1982. *Water: The Nature, Uses, and Future of Our Most Precious and Abused Resource.* New York: Farrar Straus Giroux.

Pye, Kenneth. 1987. *Aeolian Dust and Dust Deposits.* San Diego, Cal.: Academic Press.

Rambler, Mitchell B., Lynn Margulis, and Rene Fester. 1989. *Global Ecology: Towards a Science of the Biosphere.* San Diego, Cal.: Academic Press.

Redman, Charles L. 1978. *The Rise of Civilization: From Early Farmers to Urban Society in the Ancient Near East.* San Francisco: W. H. Freeman.

Regenstein, Lewis. 1982. *America the Poisoned.* Washington, D.C.: Acropolis Books.

Reij, Chris, Paul Mulder, and Louis Begemann. 1988. *Water Harvesting for Plant Production.* Washington, D.C.: World Bank (Tech. Paper No. 91).

Reisner, Marc. 1986. *Cadillac Desert.* New York: Penguin Books.

Repetto, Robert, and Malcolm Gillis, eds. 1988. *Public Policies and the Misuse of Forest Resources.* Cambridge: Cambridge University Press.

Repetto, Robert, ed. 1985. *The Global Possible: Resources, Development, and the New Century.* New Haven, Conn.: Yale University Press.

Richter, Jorg. 1987. *The Soil as a Reactor.* Cremlingen, West Germany: Catena Verlag.

Rindos, David. 1984. *The Origins of Agriculture: An Evolutionary Perspective.* San Diego, Cal.: Academic Press.

Rosenblum, Mort, and Doug Williamson. 1987. *Squandering Eden: Africa at the Edge.* San Diego, Cal.: Harcourt Brace Jovanovich.

Sagan, Dorian. 1990. *Biospheres: Metamorphosis of Planet Earth.* New York: McGraw-Hill.

Sanchez, Pedro A. 1976. *Properties and Management of Soils in the Tropics.* New York: John Wiley & Sons.

Sawhney, B. L., and K. Brown. 1989. *Reactions and Movement of Organic Chemicals in Soils.* Madison, Wis.: Soil Science Society of America.

Schioler, Thorkild. 1973. *Roman and Islamic Water-Lifting Wheels.* Copenhagen: Odense University Press.

Schwab, Glenn O., Richard K. Frevert, Talcott W. Edminster, and Kenneth K. Barnes. 1981. *Soil and Water Conservation Engineering.* New York: John Wiley & Sons.

Sears, Paul B. 1935. *Deserts on the March.* Norman, Okla.: University of Oklahoma Press.

Seymour, John, and Herbert Girardet. 1987. *Blueprint for a Green Planet.* New York: Prentice Hall.

Sheridan, David. 1981. *Desertification of the United States.* Washington, D.C.: Council on Environmental Quality, U.S. Govt. Printing Office.

Simpson-Lewis, Wendy, Ruth McKechnie, and V. Neimanis. 1983. *Stress on Land.* Ottawa: Canadian Govt. Publishing Centre.

Singer, Michael J., and Donald N. Munns. 1987. *Soils: An Introduction.* New York: Macmillan.

Southgate, Douglas. 1988. *The Economics of Land Degradation in the Third World.* Environment Department Working Paper No. 2. Washington, D.C.: World Bank.

Sparrow, H. O., ed. 1984. *Soils at Risk: Canada's Eroding Future.* Ottawa: The Senate of Canada.

Spears, John. 1988. *Containing Tropical Deforestation: A Review of Priority Areas for Technological and Policy Research.* Washington, D.C.: World Bank (Environment Department Working Paper No. 10).

Stern, Peter. 1979. *Small Scale Irrigation.* Bet Dagan, Israel: International Irrigation Information Center.

Stewart, J. Ian. 1988. *Response Farming in Rainfed Agriculture.* Davis, Cal.: Wharf Foundation Press.

Tank, Ronald W. 1983. *Environmental Geology.* Oxford: Oxford University Press.

Thirgood, J. V. 1981. *Man and the Mediterranean Forest: A History of Resource Depletion.* New York: Academic Press.

Timberlake, Lloyd. 1985. *Africa in Crisis: The Causes, the Cures of Environmental Bankruptcy*. London: International Institute for Environment and Development.

Toy, Terrence J., and Richard F. Hadley. 1987. *Geomorphology and Reclamation of Disturbed Lands*. San Diego, Cal.: Academic Press.

Turner, A. K., ed. 1984. *Soil-Water Management*. Canberra: International Development Program of Australian Universities (IDP).

Ward, C. H., W. Giger, and P. L. McCarty, eds. 1985. *Ground Water Quality*. New York: John Wiley & Sons.

Waring, Richard H., and William H. Schlesinger. 1985. *Forest Ecosystems: Concepts and Management*. San Diego, Cal.: Academic Press.

Waterhouse, James. 1982. *Water Engineering for Agriculture*. London: Batsford.

Wennergren, E. B., D. L. Plucknett, N. J. H. Smith, W. L. Furlong, and J. H. Joshi. 1986. *Solving World Hunger: The U.S. Stake*. Cabin John, Md.: Seven Locks Press.

Whalley, Joyce Irene. 1982. *Pliny the Elder, Historia Naturalis*. London: Victoria and Albert Museum.

White, R. E. 1979. *Introduction to the Principles and Practice of Soil Science*. Oxford: Blackwell.

Whittington, D., and G. Guariso. 1983. *Water Management Models in Practice: A Case Study of the Aswan High Dam*. Amsterdam: Elsevier.

Wild, Alan, ed. 1988. *Russell's Soil Conditions and Plant Growth*. Harlow, Essex, England: Longman.

Wilson, E. O., ed. 1988. *Biodiversity*. Washington, D.C.: National Academy Press.

Wolman, M. G., and F. G. A. Fournier, eds. 1987. *Land Transformation in Agriculture*. New York: John Wiley & Sons.

Worster, Donald. 1985. *Rivers of Empire: Water, Aridity, and the Growth of the American West*. New York: Pantheon Books.

Yair, Aaron, and Simon Berkowicz, eds. 1989. *Arid and SemiArid Environments*. Cremlingen-Destedt, West Germany: Catena Verlag.

INDEX